Maria Afentakis BSc, M

THE SPIRITUAL SCIENTIST

Bridging the gap between science and spirituality

THE SPIRITUAL SCIENTIST

Bridging the gap between science and spirituality

Maria Afentakis BSc, MSc

© 2020 Maria Afentakis
First published in Great Britain in 2020 by
Eclipse Publishing & Media Ltd
34 Winchester Road
Orpington
Kent
BR6 9DP

www.epmbooks.co.uk

All rights reserved. No part of this publication may be reproduced, stored in a retrieval system, or transmitted, in any form or by any means, electronic, mechanical, photocopying, recording or otherwise, without the prior written permission of the publisher. Reprographic reproduction is permitted in accordance with the terms and licences issued by the Copyright Licensing Agency.

Disclaimer

The author and publisher believe that the sources of information upon which this book is based are reliable and have made every effort to ensure the accuracy of the text. However, neither the publisher nor the author can accept any legal responsibility whatsoever for consequences that may arise from errors or omissions, or from any opinion or advice given.

This book is intended as general information only. It is not intended to replace professional medical advice, nor should it be used to diagnose or treat any health condition. For diagnosis or treatment of any medical problem, consult your own qualified medical practitioner. The publisher and author disclaim any liability arising directly or indirectly from the use of this book and are not responsible for any specific health needs that may require medical supervision and are not liable for any damages or negative consequences from any treatment, action, application or preparation, to any person reading or following the information in this book.

ISBN 978-1-912839-01-8 (paperback)
ISBN 978-1-912839-51-3 (ebook)

Cover design by Aimee Coveney
Copy-edited by Graham Hughes
Text design by Daisy Editorial
Printed in Great Britain by 4Edge Limited, Hockley, Essex, UK

Distributed by UK: Star Book Sales/Orca
                US: SCB Distributors

Contents

Endorsements	ix
Introduction	1
1. What is energy?	12
2. Energetic aura	25
3. The human body	46
4. The chakras	68
5. Grounding and balancing the chakras	100
6. Crystals and healing	111
7. Essential oils and aromatherapy	129
8. The brain and mindfulness	151
9. How to make your own meditation kit	169
An invitation to connect and engage	180
Gratitude	183
About Maria Afentakis	185

I would like to dedicate this book to my late father, Dimitris Afentakis, with love and light. He always would say to me, 'Spread your wings and fly! You can achieve anything you put your mind and heart to.'

I feel I have done you proud in writing this special book to bridge the gap between science and spirituality and to spread this knowledge to help, heal and educate the world and beyond!

I will never forget you, Dad – you will always be in my heart!

The Spiritual Scientist

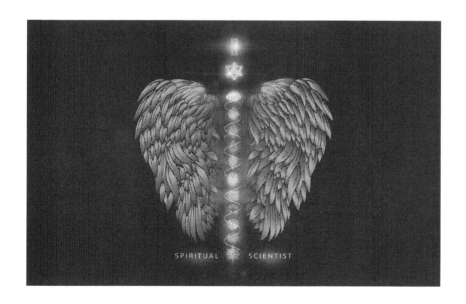

Endorsements

'Science has not always been split from spirituality – there are many examples of notable scientists who had a deep connection to the spiritual side of life. In more recent years it has been much harder finding men and women of science who are willing to embrace the spiritual dimension. Maria is one of those great pioneering spirits who are stepping forward to heal this painful and artificial schism. *The Spiritual Scientist* is a much-needed book that bridges the artificial gap between science and spirituality. Most importantly, in this time of spiritual darkness, she helps the reader understand both scientific and spiritual notions of how energy works, which in turn helps restore a sense of faith in the Universe along with a greater experience of balance and harmony within their energy field.'

– Steve Ahnael Nobel, author, book mentor, coach, founder of Soul Matrix Healing
(London, UK)

'Maria is an exceptionally gifted therapist and healer. I had the chance to experience Maria's work several times and she never stops impressing me. Through her high intuition and sensitivity she mastered her way of tuning into the subtleties of energy of places, people and animals, allowing her to guide her patients through a true path of healing and transformation. Everything in the universe, including thoughts, feelings and all forms of matter, is comprised of energy. With this book Maria takes a step forward to show us how energy, the continuous source of life, is the "missing" link between science and spirituality.'

– Vera Martins, PhD, ND, AMH and reiki master
(London, UK)

'In her book, *The Spiritual Scientist*, Maria explains the connections of basic scientific principles and spiritual concepts of healing and energy from a degreed biochemist and neurologist's point of view. Her book is unique in that she creates a bridge for the scientifically-minded person to acquire a foundational learning of spiritual energy and how to use it for healing purposes. A wonderful introduction for anyone who wishes to begin a spiritual practice of energy work.'

– Kris Seraphine-Oster, PhD, author, marketing strategist, business coach
(Santa Barbara, USA)

'As a fellow scientist, Maria Afentakis has captured the heart and soul of Albert Einstein with great effect in this fascinating and informative book. Through Einstein's ground-breaking teachings and philosophical views, Maria offers the reader two bites of the cherry with a frank scientific explanation translated into a practical, accessible and elegant way for all to understand. *The Spiritual Scientist* beautifully marries together Einstein's scientific and spiritual beliefs into the one common denominator: we are all energy.'

– Elizabeth Whiter, author of *You Can Heal Your Pet* and *The Animal Healer*
(Hassocks, UK)

'Maria Afentakis has done a brilliant job of taking her complex knowledge of science and breaking it down so that the average person can understand it. She deftly illustrates how science works using everyday examples in a manner that is easy, informative and entertaining. She gives one just enough without overwhelming!

'At the end of each chapter she brings it home by taking the scientific principle she has illustrated and relays it from a spiritual perspective, giving the reader useful exercises and examples of how to apply this scientific knowledge in a spiritual way. I'm truly amazed! I could read forever and not only learn more about science but also about spirituality!

'Maria is also very relatable. She shares her own personal experiences – giving us a peek into her world – reminding us that this brilliant scientist is also a human being on a spiritual journey, like the rest of us. If you like learning new things, and have an open mind, you will really enjoy this book!'

– Sharon Prince, international spiritual teacher, intuitive and author
(Texas, USA)

Introduction

Let me begin by asking you a few questions. Are you sensitive to the energies of certain individuals and places? Are you tired and drained? Would you like to feel balanced, happy and at ease every single day? Would you like to be free of anxiety and depression? Would you like to know how spiritual practices work by understanding the science behind them?

If your answer to any of these questions is yes, this book will be a useful guide and a great source of information. After reading it a few times, you will feel energised, balanced, tranquil and happy, no matter what life throws at you!

About the book

I am the bridge between both worlds: the spiritual and the scientific. Coming from a scientific background, I hope, by explaining the concepts of energy, I can convince you about the importance of spiritual practices and how you can achieve optimal health for your mind, body and soul by implementing these concepts in your day-to-day life.

This book is an easy-read guide to science and spirituality: both such diverse subjects, but strongly connected. In a way, it is like connecting left-brain thinking (logic, mathematics, science) with right-brain thinking (intuition, creativity, imagination). The main aim of this book is to bridge the gap between the physical concept of energy and the spiritual concept of energy.

I will take you on a journey of discovery. As you begin reading, you will gain basic scientific knowledge that will be greatly enhanced as you work though the book. It will be like crossing a bridge – at each stage of the bridge you will gain another piece of knowledge; as you reach the end of it, your bag of knowledge will turn into a sack of useful information! You will gain a greater understanding regarding different spiritual practices and how important these are in maintaining a balanced well-being.

In this book, you will explore the science behind physical energy – with particular attention to the electromagnetic spectrum, vibrations and frequencies, light energy, how the human body works and the function of the brain. The spiritual concepts of energetic aura, the chakra system, balancing and restoring energy in your aura and the importance of mindfulness techniques will be explained to increase your knowledge of the spiritual concepts of energy.

Spiritual practices include using crystals to balance and restore blocked chakras, essential oils and meditation to help with relaxation, and positive affirmations for increasing energetic vibration to connect with one's energy and the energies of earth and all living beings.

The last chapter will bring all the knowledge together with a practical step-by-step guide. You will learn how to create a variety of meditation kits using essential oils, crystals and mindfulness techniques. For example, if you are feeling tired and stressed, you could create a healing meditation kit with peppermint and lavender essential oils and pink and green crystals (refer to p171 for recipe).

10 major things you will accomplish in this book

1. Explore the meaning of energy in science
2. Understand energetic aura and how it works
3. Explore the human body and its systems
4. Recognise which chakras correspond to specific parts of the body
5. Learn techniques to balance and restore inactive or blocked chakras
6. Use crystals to heal and balance energy centres
7. Discover the use of essential oils
8. Learn about the structure and function of the brain
9. Use mindfulness techniques to overcome stress
10. Create your own meditation kit

How to use the book

I would recommend reading the book from beginning to end at least once. This is so you can understand the scientific and spiritual concepts of energy. You can refer to individual chapters where you need help on a particular practice or need more information on a specific topic.

The guided meditations, affirmations and activities throughout the book are to assist you and to ensure you fully understand each topic.

» Preparation guidance for meditations
- Sit or lie down and make sure you are rested and feeling comfortable.
- It would be advisable to cover yourself with a blanket, because your body temperature drops during meditation.
- Wear loose clothing for comfort.
- Make sure you will not be disturbed – no mobile phones, TV or any other distractions.

As this book is a guide and self-help book for balancing your energy, you will have your own unique experience. You may wish to document your experiences in a journal as you go through the book.

My story

Before we embark on a journey of discovering the science behind spirituality and spiritual practices, let me begin by explaining how I became the Spiritual Scientist.

From a young age, I was highly sensitive to energetic fields of places and people, and felt that every living thing had an energetic aura/field. My mother recalls seeing butterflies always following me, as though they were attracted to my magnetic energetic aura. I was always running away in my imagination and was fascinated by mermaids, fairies, unicorns, dragons and anything magical, and I still am to this day!

Growing up, I was always surrounded by spiritual practice. My grandmother would have coffee meetings with her friends, where she would read their fortunes in the coffee cup. My mum would have tarot card readings and crystal ball evenings, where she would give channelled messages to her friends. I would watch excitedly, eager to learn more!

When I was in my twenties, my mother bought me an angel oracle card deck by Doreen Virtue called 'Daily Guidance from Your Angels'. I enjoyed using this deck, and my interest in angels grew.

I started reading more books written by Doreen Virtue and other angel experts to broaden my knowledge. These books included topics such as archangels, elementals, fairies and mermaids. I bought oracle cards to connect to the angels and elemental beings, and gave card readings to friends and family. At this stage I had no idea I was an intuitive channel – until I began giving readings to members of the public and getting very positive feedback on their accuracy.

My path in life was beginning to take shape; my spiritual work was giving me immense pleasure, so I started to embark on courses and workshops in angelic reiki, angelic healing, animal reiki and embracing the angelic realm. During this training, I learned to connect to the angels and elemental beings by meditation, healing and visualisation.

» The incarnated angel

I guess I have always known I had a connection to the angelic realm, because I have always had a fascination with candles, icons and angel statues. I feel relaxed and at peace when visiting a church or a sacred place.

During a channelled reading given by an intuitive channel, it was confirmed that I am an *incarnated angel*. What does 'incarnated angel' actually mean? It means I am here on earth as a messenger to help others find happiness, love and peace in their lives by educating them on balancing their energy.

Here is the channelled message:

> *Maria, the angels are letting you know that you have been divinely appointed to work as a representative for the angelic kingdom. You have a strong affinity with the angels and you like working with angelic energy, but did you know you are an incarnated angel? The angels are confirming that. You need to remember that you have your own team of angels, as well as other teams that are here to help you, and you don't need to work alone. You can ask for help!*
>
> *You need to protect yourself from the public because you are overly sensitive to energy, especially people that have negative energies and are mean. They work on a very low vibration and are attracted to your angelic high vibrational light aura, so they try to invade it and make you feel sad and drain your energy.*

> *So, you need to learn to protect your energy, and remember that any incarnated angel is strongly associated with Archangel Michael, and you can ask him to protect you, help you and work with you. You can open up your energetic aura in private settings to help others – you are a teacher and mentor. You will use your skills as a scientist and your spiritual connection to the angelic realm to help you communicate this information.*
>
> Channelled by Sharon Prince, intuitive channel and guide

This message was extremely profound and deeply overwhelming, but made a lot of sense.

This information gave me the inspiration to research angel experts who have had experience of meeting incarnated angels. The first book I read was *Earth Angel Realms* by Doreen Virtue – it explained a lot about my characteristics and the experiences I have encountered.

An *earth angel*, as described by Doreen Virtue, is

> *a soul who is elected to be born into a human body to have a greater impact and make the world a better place; earth angels don't just have a soul purpose but also a global mission to provide a service to the world.*

Earth angels are powerful lightworkers with a legacy of healing and miracles behind them. There are many different types of earth angels including incarnated angels, incarnated elementals, star beings, mystic angels, cherubs and many more.

After reading the chapter on incarnated angels, I found that the channelled message about being an incarnated angel was accurate – I felt I had come home in a sense.

Incarnated angels are identified by specific characteristics including:
- having a pure love in their hearts
- putting others before themselves

- a white angelic aura which attracts animals, children and humans
- weight issues
- a beautiful heart-shaped face
- a large body frame (typically seen in pictures/paintings of angels)
- loving anything related to angels (pictures, statues, churches, books, meditations and workshops on angels)
- being talkative
- others identify them as an angel
- working in a helping profession such as healer, nurse or teacher
- apologising a lot and carrying guilt
- having a problem saying no to people
- having friends/family members with addictions
- chronic fatigue syndrome due to absorbing and taking on others' toxic energy out in public as well as in the workplace.

The book gave some guidance on protecting my energetic aura and being more assertive. Here it is:
- Incarnated angels must learn to protect and shield their energy and light.
- Wait before saying yes, and say something like 'I'll get back to you on this'.
- Replace an apology with a positive affirmation.
- Engage in cardiovascular exercise to keep stress levels and weight under control.
- Balance giving and receiving.
- Drop any rules that are restricting you from your soul and global mission.
- Ask people for help if needed.
- Allow others to give you compliments.
- Most importantly, spread your wings and shine your light to the world.

The description of an incarnated angel truly fits my characteristics (body shape and facial features), personality and job. It was great learning how to protect my energy and feel better so I can be on track with my global and soul mission.

During my training to become an angelic reiki practitioner, a common recurrence was evident: I was always protecting and healing the heart chakra to bring lots of love and light to the recipient. In one healing session, I felt pain in my back, as if my wings were trying to break through my skin – this feeling of my wings breaking out is a common occurrence.

On another special occasion, one morning on my way to work, I was meditating to some calm, relaxing music, I closed my eyes and saw my white feathered wings opening out; I felt deeply emotional and wanted to go back home to the kingdom of angels.

Later that day, I went to a Mary Magdalene meditation and received a healing and channelled message, again from the same lady that gave me the information about being an incarnated angel.

The channelled message I received was:

> *Dearest one,*
> *You have suffered a lot in this lifetime because of others, who are mean to you and exert their negative energies and personalities on to you. You are an incarnated angel working on Archangel Michael's team of indigo warrior angels. Do not worry! I love you so much and you are doing a great job. You will be a healer full-time and have your own spiritual practice.*
>
> Channelled by Sharon Prince, intuitive channel and guide

It was truly enlightening receiving the love and healing. Even though I sometimes find it hard and challenging on earth, I know I am here on a soul mission to bring love and light to this planet, heal animals and humans, teach others so they can understand the importance of looking after themselves and their energetic aura, and finally to communicate to and inspire the whole planet through my books, talks and workshops.

» The scientist

During my years at high school, my passion lay in the scientific subjects, including Biology, Chemistry, Physics and Psychology. All were very complex areas, but I was determined to learn and study these subjects in detail. They were of interest to me because I wanted to learn how the world worked, how the mind worked, and about life and energy.

I studied A levels in Biology, Chemistry and Psychology and went on to complete a BSc (Hons) in Biochemistry and an MSc in Neuroscience at university. Again, both very complex subjects. This intense study taught me a lot, including learning to be patient, determined, focused and able to achieve everything I wanted to academically.

I worked as a scientific researcher in the field of breast cancer research at The Royal Marsden Hospital for just over 10 years. My work involved managing clinical trials and investigating individualised treatments for patients with breast cancer. The practical work involved looking at breast cancer tumours under a light microscope and diagnosing what types of tumour they were. Depending on the type, the clinical oncologist would give the patient a specific drug.

As well as looking at the histology of the breast tumour, I extracted DNA and RNA (ribonucleic acid – essential for all forms of life) from it to be sent for genetic analysis, to identify mutated and inherited genes in the patient's cancer tissue.

In 2011, I was given my own research project alongside my general work. This project involved investigating a protein called *BAG1* which potentially could have an impact on diagnosis, treatment and disease reoccurrence in breast cancer patients.

The project involved using a clinical trial set of patients where the clinical, genetic and scientific data was available. I used a technique called *immunohistochemistry*, which involves looking at a section of the tumour stained with BAG1 antibody and assessing the expression using a light microscope. It does sound a bit complicated, but this is the easiest explanation. The reference for my paper is in the biography ('About Maria Afentakis BSc, MSc') at the end of the book, if you are brave enough to read it!

The research was a great success, and I produced a first author publication in a highly prestigious breast cancer research journal. I presented my findings at the biggest breast cancer conference in the world in San Antonio, Texas. This was by far the greatest achievement in my scientific career – it gave me confidence in communicating scientific information and explaining my work to all types of people, from junior technicians to the principal investigators / heads of department in scientific research establishments.

I am now working as a senior research scientist at Barts Cancer Institute, in the Department of Tumour Biology. My role involves providing a service to the whole institute and other cancer research institutes in all areas of the histopathology of cancer.

All my scientific knowledge and experience has been invaluable to my career, and I feel it is important in the work I plan to do now as an author. This concludes my story and how the Spiritual Scientist was born. I am here to bridge the gap between science and spirituality by educating people on the scientific and spiritual concepts of energy to have a relaxed, balanced and happy life. I hope you enjoy reading this book and, whether you are a healer, an interested reader, a therapist or an academic, find something of value within its pages.

Reading

Virtue D. *Earth Angel Realms: Revised and Updated Information for Incarnated Angels, Elementals, Wizards and Other Lightworkers*. Hay House, 2014.

CHAPTER 1

What is energy?

I would like to begin this chapter with the scientist who has inspired me to write about the physics of energy: Albert Einstein – the most famous physicist who ever lived. He was extremely intelligent and was not scared to put his theories and knowledge to the test no matter what! He struggled through school, as he wasn't good in most subjects – apart from maths and physics, where he was exceptionally talented. He was a great inspiration to the world. He created the most famous and relevant equation in energy! That was the general theory of relativity:

$$E = mc^2$$

E – energy (measured in joules, j)

m – mass (measured in kg)

c – speed of light (300 million m/s)

This equation indicates that energy and mass are the entity and can be changed from one into the other. Energy is equal to mass multiplied by the speed of light squared. The equation resembles the body at rest, which indicates that a modest amount of mass is equal to the energy needed for the body to function. Another relevant meaning of this equation is that mass and energy are the same as conserved physical quantity.

Albert Einstein won the prestigious Nobel Prize in Physics in 1921, not for his work on the theory of relativity, but for his work on the photoelectric effect. This is a well-known phenomenon: metal emits electrons when light shines upon them. He developed quantum theory – the theoretical basis of modern physics that explains the nature and behaviour of matter and energy on the atomic and subatomic level. In 1917, he applied the general theory of relativity to model the structure of the universe. What a remarkable scientist! Throughout his life he published over 300 scientific papers and 150 non-scientific ones (mainly philosophical findings).

Another interesting fact is that after his death, his brain was stolen by the pathologist, who preserved it in jars for the next 40 years. Finally, he decided to give Einstein's brain to neuroscientists to investigate it. They found that the parietal lobes were 20 per cent larger than in a normal human brain. The parietal lobes are responsible for mathematical, visual and spatial cognition, which explained why his ability in these areas was so enhanced. Further investigations showed his brain contained 17 per cent more neurons than a normal human brain, which indicates more neuron–neuron connections occurring, more synapses reached and a greater amount of brain activity. This could explain why he was super-intelligent from a young age. It was recorded that he taught himself different aspects of mathematics including algebra and geometry.

He didn't just have exceptional mathematical skills, but also an amazing imagination. He had very profound views on life and created over 100 notable quotes on nature, life, philosophy, science, religion, love, imagination, the world and politics. A book recommendation with his top quotes is at the end of this chapter. He used both his scientific knowledge and his spiritual skills to educate the world about how energy works. This is a great similarity I share with Albert Einstein.

Einstein's work wasn't just about the theories of physics, but also on the theories of philosophy: I guess now it's called 'spirituality'.

Albert Einstein was the spiritual scientist, and this is the reason why I felt I needed to learn and connect with him – as I am the modern-day Spiritual Scientist!

Albert Einstein first came into my thoughts when I was having a conversation with Elizabeth Whiter, a well-known animal healer. She suggested that I should connect with him, and this inspired me to do so and to write about him in my book. A few days later, I was staying at an Airbnb where the resident, a talented artist called Helen Lunt, had painted a magnificent portrait of Einstein. As soon as I saw it, I had to buy it. The artist channelled information about Albert Einstein, and it is represented in her portrait. It was so inspirational! It was a certainly another sign that I needed to connect with him. I have the portrait now above my writing desk for inspiration (refer to Figure 1.1).

The channelled information and her description of the portrait is as follows:

> *The title is 'Intelligence Is Having Fun!' I loved the picture of him pulling his tongue out, so I drew it. The sun going in through his head and out on his tongue is the kinetic energy of the body's movement. On his scarf are the quotes that I liked.*
>
> *There are planets at the top of the portrait, and towards the bottom left are two flowers representing reproduction, sexuality and life. Top-left-hand is the earth's crust cut in half. The bee represents life, and we will die without them. Healing hands in bottom right, a butterfly and the bat represent transformation.*
>
> Helen Lunt, aspiring artist

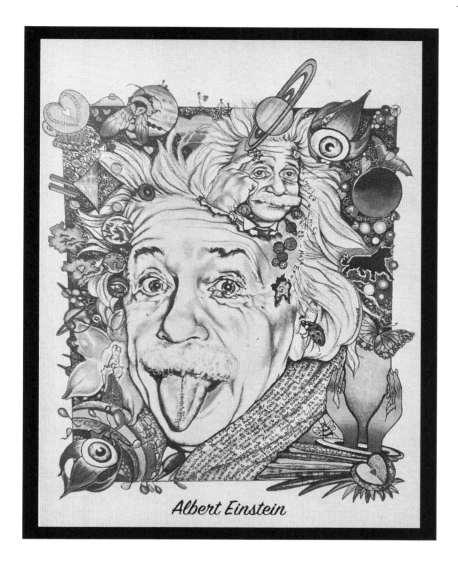

Figure 1.1: Portrait of Albert Einstein – painted by Helen Lunt ©

Einstein's quotes on the scarf are very insightful and inspirational:
- 'Logic will get you from A–B, imagination will take you everywhere.' (Very true.)

- 'Anyone who has never made a mistake has never tried anything new.'
- 'Try not to become a man of success but a man of value.'
- 'Life is like riding a bike. To keep your balance, you must keep moving.' (Exactly.)
- 'Coincidence is God's way of remaining anonymous.'
- 'The only reason for time is so that everything doesn't happen at once.'
- 'The most beautiful thing we can experience is the mysterious. It is the source of all true art and all science.' (This is my favourite one.)
- 'Great spirits have always encountered violent opposition from mediocre minds.' (Definitely some truth in this.)

There are a few other inspirational quotes I wanted to share:
- 'Imagination is more important than knowledge. Knowledge is limited. Imagination encircles the world.'
- 'Look deeper into nature and then you will understand everything better.'
- 'All religions, arts and sciences are branches of the same tree.'
- 'Education is not the learning of facts, but the training of the mind to think.'
- 'It is the supreme art of the teacher to awaken joy in creative expression and knowledge.'
- 'The only source of knowledge is experience.'

As you can see, his words were born of intelligence and inspiration and made a lot of sense.

On 21 January 2019, it was the first full moon of the year. I received a channelled message from Albert Einstein at 3am. I remember this night so vividly. I had just had a relaxing full-moon bath with salts and oils – I felt so peaceful but yet so energetic! I was sitting by the window in my living room. I was admiring the amazingly bright full moon. It was a special blood-red moon. Here is the channelled message I received:

You must start as you mean to go on! You are the bridge between the science world and the spiritual world! You are here on a very special mission! I am here to help and guide you in any way possible. You will be a star – you are magnificent! You will have an extremely busy schedule, but I know you can handle it! But please remember to look after yourself – nourish, self-love, laugh, be happy, feel joy and, most importantly, be with people that are loving towards you! You have built up your strength over the past few months, so you are powerful now!

Now, you must make a plan for your first book; there will be many more! Stick to the plan, write and write every day! You can do it, my dear! You are the modern-day Spiritual Scientist! You are so intelligent, yet so spiritual. Healing the animals will give you lots of love and comfort. This is needed for you on your magical journey! Please look after yourself and if you need anything, I am here for you! Now, please go to sleep and rest. Lots of energetic and vibrant love, Albert – please call me Albert from now on!

This message was so heart-warming, and it made me feel happy to connect with Albert Einstein and write this book to help others understand the physical concepts of energy and how to keep their energies balanced by incorporating spiritual practices into their everyday lives.

Now, I will begin explaining what energy is and the different types of energy, with particular attention to chemical energy, which is important for plant, human and animal survival; and to visible light in the electromagnetic spectrum. Energy is vital for every living organism and system in biology, chemistry, physics, the environment and geology. Each organism or system receives energy by different means: for example, humans get energy from food, plants get energy by photosynthesising, cars get energy from fuel and the earth gets energy from the sun.

The word 'energy' is derived from the Greek word *energia* (ἐνέργεια). In ancient Greece, ἐνέργεια had a much broader meaning than it does now – it wasn't just referred to as a physical concept, but also extended to philosophical concepts such as happiness, imagination and pleasure. Physical energy is a quantitative measure which can be transferred to an object to either move or heat up the object. The energy law of conservation states:

> *Energy cannot be made or destroyed, only transferred from one form to another.*

For example, a coal fire will start burning by the process of combustion, which begins as chemical energy and then converts to heat and light energy. The energy has not been destroyed – only transformed to another type or types. There are many types of energy including kinetic (movement), potential (work done), gravitational potential, chemical, light, sound and thermal energy. Kinetic, chemical and light energy will be discussed in this chapter.

Kinetic energy is when an object is moving. The kinetic energy of a moving object is dependent on its mass and its speed, as illustrated in the equation below:

$$E \text{ (energy)} = 0.5 \times M \text{ (mass)} \times V^2 \text{ (volume)}$$

An example of kinetic energy is a moving car. The lighter its mass, the faster the car will move. Another example of kinetic energy is the three states of matter: gas, liquid and solid. In a gas, the particles have a lot of kinetic energy and are free to move in any direction. In a liquid, the particles have a bit more kinetic energy and fill up the space of the container. In a solid, the particles are much more compact and there is very little movement between them, therefore less kinetic energy will occur (refer to Figure 1.2 for a visual representation of the three states of matter).

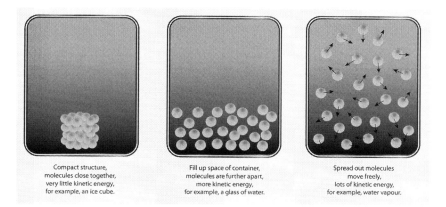

Compact structure, molecules close together, very little kinetic energy, for example, an ice cube.

Fill up space of container, molecules are further apart, more kinetic energy, for example, a glass of water.

Spread out molecules move freely, lots of kinetic energy, for example, water vapour.

Figure 1.2: The kinetic energy of the three states of matter in water: solid, liquid and gas

Chemical energy is when two or more types of compounds of different elements combine to form products. An element is a substance that is made up of only one type of atom. An atom is the smallest, basic form of matter. The periodic table contains all the elements with their number of atoms – for example, hydrogen is at position 1 in the periodic table and consists of one atom. A compound is the combination of one or more elements.

An example of chemical energy is a process called *photosynthesis* (refer to Figure 1.3). This is how plants make food for survival. Photosynthesis involves different chemical reactions. These reactions convert energy from one form into another. Trees and plants have leaves, which have a major component in their cells called *chloroplasts*. Chloroplasts are tiny compartments made up of membranes, which contain a green pigment called *chlorophyll*.

This is the pigment that absorbs light energy from the sun to make food (glucose – a carbohydrate). The greener the leaves of a plant, the more photosynthesis can occur.

The equation for photosynthesis is:

carbon dioxide + water and light energy = glucose and oxygen

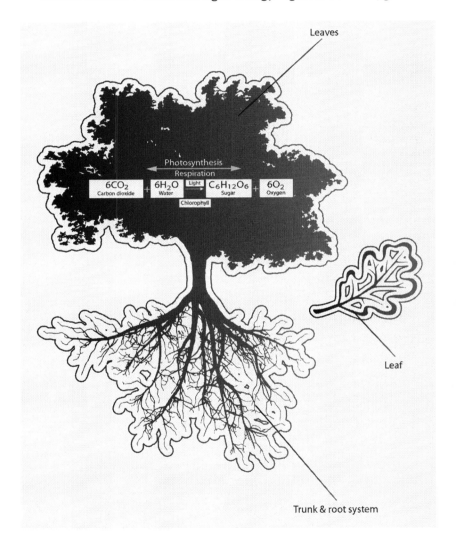

Figure 1.3: An example of chemical energy: the process of photosynthesis

The most important concepts to remember regarding chemical reactions are that all elements of each compound must be balanced on both sides of the formula, and total energy will always remain the same even after it is transformed into another type. Examples of elements include carbon (C), nitrogen (N) and oxygen (O). A compound is a substance which contains two or more elements. If you reverse the photosynthesis equation, the process is called *respiration*.

The equation for respiration is:

glucose and oxygen = carbon dioxide + water and ATP (energy)

It is the most important process for all living cells in the human body. Respiration is the production of energy through the intake of oxygen and the release of carbon dioxide by oxidisation of a carbon compound (glucose). It takes place within all cells. The energy molecules that are produced are called *adenosine triphosphate* (ATP). ATP is in the *mitochondrion organelle*, which is located in the cytoplasm of every single cell in the body. (The structure and functions of every organelle within a cell will be discussed in Chapter 3: 'The human body'.)

Cellular respiration can be aerobic (with oxygen) – such as aerobic exercise, where oxygen is needed, or anaerobic exercise (without oxygen), where lactic acid is used instead of oxygen. To explain this further, think of long-distance runners: once all their oxygen has been used, it is replaced by lactic acid. This is the reason why marathon runners or people who do a lot of strenuous exercise are prone to getting cramps in the legs and feet. The cramping is due to build-up of lactic acid.

The three main food groups needed for the body to function and be healthy are:
- **proteins** – vital for the structure component of many body parts such as hair and nails, and most chemical reactions in the body
- **carbohydrates** – the main source of energy to help you with daily activities as well as exercise

- **fats** – needed for cells, as they make the cell membrane, which is the protective barrier of the cell and allows certain smaller molecules into it.

To be fit and healthy, a well-balanced diet with all these food groups as well as fruit and vegetables is really important, so the body can function in the best way possible.

Light energy is a type of electromagnetic radiation forming part of the electromagnetic spectrum. It is referred to as *visible light* and can be seen by the naked eye. The waves in the electromagnetic spectrum have different wavelengths and frequencies and have common properties: for example, every type of wave is able to travel through a vacuum; waves carry energy from one place to another; waves travel at the speed of light (300 million m/s); and waves can be reflected, refracted and diffracted.

Visible light has wavelengths in the range of 400 to 700 nanometres with a frequency range of 430 to 750 terahertz. The frequency is what gives the light a specific colour.

A prism can be used to demonstrate the wavelengths from the visible light spectrum. A prism is a 3D triangular shape made from glass. When it is placed into direct light, it is possible to see a spectrum of colours by a process called *diffraction*. Each colour in the electromagnetic spectrum has a unique wavelength: red light has the longest wavelength, whereas violet light has the shortest. The shorter the wavelength, the longer it takes for the light to travel through the prism.

A rainbow is another good example to explain the visible light spectrum and the different colours. When it is raining and the sun is shining, light (from the sun) is being reflected, refracted and dispersed onto raindrops, which is why a rainbow appears in the sky. The point where the rainbow appears is directly opposite the sun. A wavelength (λ) is defined as the distance between a point on one wave and the same point on another wave. The frequency (f) is defined as the number of waves produced by the source each second. The speed of a wavelength can be calculated as frequency (*f*) multiplied by wavelength (λ).

Energy can be transmitted through space or even through a material. Every individual human, animal and plant has a unique energy! The energetic vibration can change during energy transfer – for example, if you are feeling sad and depressed, your energy is at a lower vibration; to feel happy and positive, you must raise your energetic vibration to a higher frequency. Raising your energetic vibration will be discussed throughout the book by using different methods, and the next chapter will explain, in detail, the energetic aura of a human.

Affirmation

'I am energy! I am life! I am present! I am water, earth, wind and fire! I will connect with all forms of nature and animals! I am grateful for life on earth! I am tranquil and energised!'

Activity: *Explore energy in nature*
- Go out into nature, to a place where the four elements – water, fire, air and earth – meet. For example, a forest where there is a river or lake with lots of trees, and the sun is shining.
- Sit under a tree that attracts you, remove your shoes, close your eyes and connect to the energy of the tree by placing your feet flat on the ground.
- Start to meditate by imagining that at the bottom of your feet there are long tree roots going down deep into the earth and reaching the core. The roots are held down by a large ruby crystal.
- This process is called *grounding your energy*, so you can begin to meditate.

- Connect to your environment, the trees, the flowers, the animals and all other living things, and record how you feel.
- What types of energies can you feel or see?
- Say the affirmation above three times.
- Journal your experience.

Reading

Honeysett I, Tear C, Dwyer J, Poole E. *Revise GCSE Science: Complete Study and Revision Guide*. Letts Educational, 2011.

Isaacson W. *Einstein: His Life and Universe*. Pocket Books, 2008.

Muir H. *Science in Seconds: 200 Key Concepts Explained in a Picture*. Quercus Editions Ltd, 2011.

Sapiens Hub. *Albert Einstein: Collection of Quotes*. Amazon Fulfilment, 2018.

Wingate P, Gifford C, Treays R. *Essential Science*. Usborne Publishing, 1992.

CHAPTER 2

Energetic aura

This chapter has been the most difficult to complete. As I was in the process of completing it, a tragic event happened in my life: my father died unexpectedly at the age of 61. He was my guide, counsellor, supporter and, most importantly, the person I would turn to when I was feeling sad. He would calm me down in stressful situations; he encouraged me to achieve anything I wanted to and made me feel happy when I was sad. I have decided to dedicate this book to my father, Dimitris, for all his love, support and guidance throughout my life.

Two days before my father's death, I was visiting Broadstairs. I was sitting by the sea thinking about all the energetic auras I could see, feel, hear and smell: the waves of the sea crashing onto the sandy shore, the seagulls making their bird noise, the heat of the sun beaming on my face, and the excitement of children and dogs splashing around in the sea. It was such a beautiful day!

As I turned my attention to the sea, in the distance I could see a magnificent energetic aura made up of deep blue and green shades. I began to walk on the sand; I could feel the individual grains of sand exfoliating the skin underneath my feet. I could feel the energy of the sea creatures, the plants and the shells. The smell of the sea, seaweed and salt felt so energising. The sea is the oldest source of energy, consisting of ancient life forms. It is so important to look after the oceans of Planet Earth: do not dispose of litter or plastic into the sea – it is harmful to all living things.

The composition of water is H_2O – every molecule of water contains two hydrogen atoms and one oxygen atom. Our bodies are made up of 70 per cent water; hydrogen and oxygen are essential for the survival of every living cell in the body and the function of all bodily systems (the human body will be discussed in great detail in Chapter 3). I proceeded to go for a long walk along the seashore; the more I experienced the individual energies of the seafront, the more balanced and relaxed I felt. As I looked up into the sky, it was a pale blue colour with a few clouds. I saw this huge image of an angel and felt at peace.

Energetic aura

'Aura' comes from the Greek word meaning 'breath' or 'air'. An aura is described as a type of universal energy field. Energetic aura was called *prana* in ancient India and *chi* in China. This type of energy was described in many ancient civilisations. For thousands of years, people believed that any living thing – including humans, animals, plants and water – had a unique aura. This aura can change in different circumstances – for example, the energetic aura of the ocean can change from being calm to being aggressive. This is dependent upon weather conditions, seasons, the moon and the sun.

To understand energetic aura, imagine that around your whole body there are different layers of energetic fields expanding outwards. The layers are represented as different colours. The colour of layer 2 – the emotional layer – is impacted by the mental state of the living being, so depending on how you are feeling at the time, the colour can change (see Table 2.1). This is similar to what we see in the electromagnetic spectrum of visible light, whereby different colours seen by the naked eye have their own unique wavelength and frequency. External influences such as trauma and drug use can alter energetic aura.

Some people have the ability to see energetic auras of humans and animals, or even spiritual beings such as angels. The more you understand and connect with your own energetic aura, the more likely it is that you will be able to see energetic auras of other living beings.

There are people who claim they can read one's energetic aura by using a special camera, and each specific colour has a different meaning. An aura camera works when the individual places their hands on a metal plate in front of the camera; a head-and-shoulder image is taken showing that person's predominant aura colour. The colour indicates how the person is feeling at the time of having the photo taken. Aura photos work by connecting to the energetic magnetic field of the person's aura to produce the specific colour. More than one colour can be detected and will be shown in the photo. Colours can be seen around various areas of the body.

The energetic aura consists of seven layers/fields:

1. **The etheric layer** is the first layer, closest to the physical body. It is associated with the five senses: touch, smell, sound, taste and sight. The energetic aura colour for this layer is red, and it is associated with grounding, basic and survival instincts.
2. **The emotional layer** is associated with all the emotions that radiate through the physical body; this can be seen as different colours of light, similar to visible light in the electromagnetic spectrum. The colour of this layer is dependent on the person's emotion at any given time. This layer is associated with creativity and sexuality.
3. **The mental body** is similar to layer 1, although the focus is on the brain and the mind. This layer has a bright yellow colour and governs mental activities such as thoughts, ideas, knowledge, power and intelligence.

4. **The astral body** is a bright green colour with specks of pink. This layer governs the ability to feel love, compassion and empathy to others and oneself by connecting with the energetic aura of animals and humans.
5. **The etheric template body layer** is a beautiful blue colour. It is associated with divine will, expression of one's truth, development of ideas, and all forms of communication: writing, listening, speaking and downloading information from source.
6. **The celestial body** is an indigo / deep blue chakra. It is associated with intuition, imagination, sleeping, inspiration, fears and phobias. It is like a door connecting the spiritual self to the spiritual realm.
7. **The causal body** is a violet colour and merges into gold and white light. It is the connection with divine wisdom, spirituality and the physical, emotional, psychological and spiritual self.

Refer to Figure 2.1 for the seven layers of the energetic aura.

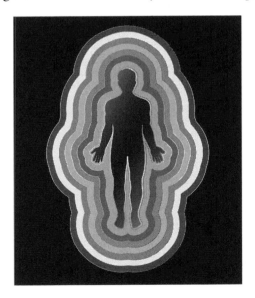

Figure 2.1: The energetic aura

Negative thoughts and actions have a profound impact on one's energetic aura. This is why it is extremely important to understand your energetic aura and be able to identify disease or illness before it manifests into the physical plane.

Colours of auras and their meanings

Table 2.1: Colours of auras and their meanings

Colour	Meaning
Aqua	Healing, compassion, empathy, sensitivity
Black	Health issues, resentment, trauma
Blue	Calm, caring, sensitive, tranquil, loving, emotional
Brown	Grounded, nature, hard-working, earth based
Gold	Enlightenment, divine guidance, protection, higher consciousness
Green	Good health, balanced, healing, friendly, kind, positivity, abundance
Indigo	Strong psychic ability, independent, honest, powerful
Orange	Creative, artistic, outgoing, vibrant, energetic, adventurous, dominant
Pink	Unconditional love, gentleness, loving, healing, helpful, compassion
Purple	Leader, intuitive, visionary, wise, connected to spirituality
Red	Passion, strong-willed, fiery, powerful, enthusiastic, determined
White	Pure, love, higher dimension
Yellow	Intelligent, happy, child-like, spiritual, imaginative, mental sphere

Energetic auras can be in a range of colours. Each colour has a unique vibration, frequency and meaning (see Table 2.1). The colour is reflective of the individual's personality at a given time, and can change when they are experiencing different feelings such as pain, jealousy, love, joy, happiness and sadness. A person's favourite colour, or the colour they are mostly drawn to, can reflect their general energetic aura.

Functions of the energetic aura

- Protects the emotional and physical energies of one's environment by screening harmful energy.
- Connects you to nourishing energies in your environment.
- Acts as a bridge to other energies on different levels of the aura.
- Holds interrelated fields that are the blueprint for your physical body, emotions, awareness and development.
- Helps you connect to others' energetic auras, such as friends, family and pets.

It is possible to connect to auras of humans, animals, nature and non-living things such as buildings and churches. This can be achieved by being connected to layers 6 and 7 of your energetic aura. It will help you open up your sensing abilities and help you see, sense, hear, smell or spiritually connect with these auras.

There are eight different types of psychic sensitivities in association with the senses:

1. **Clairvoyance** – 'clear vision'. A clairvoyant has the ability to receive information through their third-eye chakra. Information includes future (precognition), and clarifying and illuminating the present and the past (postcognition). This is received in the alpha state of the mind or during meditation. A clairvoyant may use tarot or oracle cards to provide a reading.
2. **Claircognizance** – 'clear knowing'. The ability to just know about a situation without seeing, hearing or feeling it, and responding to it emotionally or physically.

3. **Clairaudience** – 'clear audio'. The ability to perceive words or hear noises from the spiritual realm. This is achieved through the inner ear via meditation or in the alpha state of mind. The ability to hear natural sounds and noises will help with this ability.
4. **Clairempathy** – 'clear emotion'. An empath has the ability to feel other people's emotions, thoughts and symptoms by extrasensory perception. An empath must make sure they protect, restore and balance their energetic aura on a regular basis to avoid fatigue and unwanted energy.
5. **Clairesentience** – 'clear sensing'. A clairsentient perceives information by experiencing the actual emotion or physical symptom in their body. They can have a 'gut feeling' about a situation that is wrong in the other person.
(An empath feels the emotion, and a clairsentient experiences the emotion.)
6. **Clairalience** – 'clear smelling'. The ability to perceive insights by the perception of smell. This can be the smell of a place, person or animal not in their surroundings.
7. **Clairtangency** – 'clear touch', also known as *psychometry*. The ability to touch an object and perceive extrasensory information about the object by psychic channelling.
(Channelling – A channel has the ability to use their body and mind as a mechanism to connect to the spiritual realm to bring psychic information or healing to the individual.)
8. **Clairgustance** – 'clear tasting'. The ability to perceive psychic information by tasting a type of food or drink without actually consuming it.

Energetic aura of your pets

If you have a pet such as a cat, dog, rabbit, hamster or bird, they will sense your energetic aura and when it changes. For example, if you are sad or upset, your pet will try to comfort you to make you feel better. My cat will sleep next to me the whole day if I am feeling unwell or sad. Your human energy field (HEF) will combine with your pet's energy field, making one electromagnetic field, and this can bring comfort and healing to both. This is the main reason why pet owners have such a strong bond with their pets: if something happens to their pet, they feel extreme sadness and pain.

Animals work on the frequency of love and unconditional love. They tend to have a very calm and nurturing aura. They become attached to their guardian's aura and can feel their energy, as well as knowing when their guardian is leaving home to go to work or even to go abroad; they will sometimes misbehave and can become sad. It is always good to tell your pets about your planned actions, especially if you will be away or will return home late.

Energetic aura of nature

The best thing one can do is to get out in nature as much as possible: be it the seashore, the woods, the park, a river or a lake. Sit there in silence, close your eyes, spend some time listening to the sounds of the animals and nature, and see how this makes you feel. Pay attention to all the different smells. Being one with nature makes me feel amazing; working and living in London can make me feel lethargic and down, which is why I love being in nature and away from the hustle and bustle. It makes me feel revitalised, rested, balanced and peaceful.

There are many wonderful parks in London and near where I live, and it is easy to find a nice spot to connect to all of the four elements: water, air, earth and fire. Each element is important for growth and can help in different ways – for example, when I feel upset or am going through a difficult time, I like to take a trip to the seaside, where I connect to the water element (ocean). This makes me feel calm, and I can release my emotions to the sea. If I am feeling unbalanced and my energy is not stable, I go to my local park and connect to energies of the earth and the trees. This really helps ground my energy.

» The four elements and their properties

1. **Earth**: trees, woods, soil, flowers, plants and shrubs – grounding
2. **Water**: rivers, lakes, seas, waterfalls and oceans – emotions
3. **Air**: wind – communication
4. **Fire**: sun – passion and creativity

Energetic aura of water

The healing qualities of water are remarkable. I visit the white and red springs in Glastonbury three or four times a year to experience this amazing energy. Legend has it that King Arthur would regularly bathe in the waters to help heal his body. The white spring is in a cave with a few pools of cold spring water. It is said to hold the feminine energy, and I have bathed in this pool; it made me feel healed and revitalised. The red spring is in the magnificent Chalice Well gardens, and the water is ice cold and contains iron deposits. It is safe to drink, but only in small quantities; the water from the white spring can be consumed as much as it is needed. The red spring is said to hold masculine energies. Many believe it is good to balance one's masculine and feminine energies by having a glass of water containing both red and white spring water. I highly recommend a visit to the Chalice Well gardens.

Water has a memory of life and is needed for the survival of all living things. The human body is made up of 70 per cent water. For the billions of cells in the body to function properly, water is the vital ingredient. Imagine a life without water – well, don't, because life would not exist! The human body can go without food for a relatively long time, but after a few days without water, it becomes tired and dehydrated, and cannot function properly. Without water, the cells will be starved; this can have a massive impact on body tissue, organs and systems. When you connect to the element of water, it has a unique energetic aura which is soothing and can nourish the mind, body and soul.

Energetic aura of family members in spirit

For a month after the death of my father, I was feeling his presence and energy. His energy brought a sense of peace. His aura helped me in so many ways. I learned to communicate with my father in spirit through meditation as well as looking at his photos and holding his things. If you are open to this, it is possible to feel and sense the energetic aura of family members who are now in spirit form, and you may even be able to communicate with them.

Energetic aura of the moon

The moon is an astronomical body that orbits the earth in a cycle and is the earth's only natural satellite; it is seen all over the world and is the brightest object in the night sky. It is the only place humans have set foot other than the earth. The moon is around 240,000 miles from the earth and is said to be about the same age as the earth – around 4.5 billion years old. The moon originated from debris thrown into the universe, when Planet Earth hit Mars. The moon goes around the earth, as the earth goes around the sun. It takes the moon about 30 days to complete its orbit of the earth. The moon goes through a cycle called the lunar cycle, which includes the new moon, the waxing moon (meaning the moon is getting bigger every day), the full moon and the waning moon (meaning it is getting smaller every night). The energetic aura of the full moon can be extremely powerful, and it is possible to connect with her energy. I enjoy having a full-moon bath, soaking up all the luminous energy of the moon with lovely Epsom salts and essential oils. While I bathe, I think about the month that has passed: what I have achieved, what emotions I want to get rid of and who I want to forgive that month; it can be extremely therapeutic and energising. The new moon is a time for new beginnings and adventures; it is a time of manifesting what you want in the coming month.

The lunar cycle and the menstrual cycle in the female are very similar in length, and many women connect with the moon throughout their cycles. The energetic characteristics of the moon include emotional, instinctive, feminine, nurturing, safety, nourishment, mothering, grounding, childhood and homely. Connecting to the moon can help you feel these energies more intensively.

Energetic aura of the sun

The sun is the biggest star in the solar system. It is 100 times the width of the earth and could fit into all the eight planets six times. It is called a yellow dwarf star. The sun was formed 4.6 billion years ago, when a cloud of dust and gas called a nebula collapsed under its own gravity. Its mass is 330,000 times that of the earth. It accounts for 99.86 per cent of the total mass of the solar system. The sun's gravity holds the solar system together, from the biggest planets to the smallest particles of debris. It helps all eight planets, dwarf planets and the moon to be in synchronicity with each other. The connection of the sun and earth helps govern the seasons, the ocean currents, the weather, the climate and radiation. The extreme heat of the core (mid-region of the sun) fuses hydrogen atoms together to produce helium by thermo-nuclear fusion; this releases energy in the form of radiation, electricity and solar wind.

The sun is the most important source of energy for life on earth. For example, the energy from the sun is essential in the process of photosynthesis, as described in Chapter 1. We need sunlight for health and vitality. The body produces vitamin D when exposed to sunlight, and this is needed for the absorption of calcium; lack of this vitamin can lead to disorders such as rickets, depression and cognitive impairment.

A solar deity is a goddess or god who represents the sun/light. Many ancient civilisations have solar deities. For example, in Greek mythology, the sun god, Apollo, is the god of light, healing, music and prophecy. Helios, whose name means 'sun' in Greek, is the titan god of sun and sight. The goddess Alectrona (daughter of Helios) is the goddess of the morning; and Athena is the Greek goddess of wisdom and crafts, with solar deity characteristics, which makes a lot of sense: every time I have visited the temple of Athena and the ancient ruins in the Acropolis in Athens, it has always been extremely sunny, bright and hot; I have always felt her energetic aura was present in the form of the sun. The city of Athens was named after her.

Generations of cultures had so much respect and appreciation for the sun, and would have worship ceremonies. The sun is essential in bringing light and heat, which helps plants and trees to grow and lights up the world. To connect with the sun, it would be good to do a meditation at sunrise to welcome it and bring in the day, and at sunset to say goodbye to the sun and bring in the night, the stars and the moon.

Balancing energetic aura

Holistic techniques such as reiki and acupuncture can restore an individual's aura back to being calm and relaxed after a traumatic experience such as abuse, ill health or depression. There are many scientific studies that prove the benefit of using holistic techniques to treat a number of conditions, including back pain, headaches, depression, anxiety and weight loss. The holistic techniques discussed below include acupuncture, reiki and animal reiki.

» Acupuncture

Acupuncture is the fastest-growing complementary medicine in the west, and is used in combination with modern medicine. This practice has been effective in treating many diseases. In Chinese medicine, it is used alongside herbs, exercise and diet to treat diseases. The word 'acupuncture' comes from the Latin word *acus*, meaning 'needle': it means to puncture with a needle.

Acupuncture is believed to have originated in China around 100 BCE. During a treatment, the acupuncturist inserts needles into certain skin points on the body, called meridians, to assist with the flow of energy known as *qi*. Qi governs the balance of all body systems. The main functions of the universal energy qi include involuntary and voluntary movements, providing a mind and body connection, providing strong immunity, keeping blood at a stable temperature and keeping all the organs in their correct positions in the body. The treatment lasts for up to 20 minutes; sometimes heat, pressure or laser light is applied. This treatment is classed as *non-invasive*. The needle combines with the electromagnetic field of that part of the body and will restore balance, remove any blockages and create a more positive energetic field. To become an acupuncturist, it takes years of study and practice to place the needles and treat the patient accurately, with care and sensitivity. If you would like to have a treatment in the UK, you must choose a professional practitioner listed by the British Acupuncture Council.

In 2001, the World Health Organization (WHO) reviewed many trials where acupuncture was used as an alternative treatment, and concluded that it was beneficial in 28 conditions, including post-cancer treatment such as chemotherapy or radiotherapy; depression; period pain; hypertension; headaches; infertility; strokes; and pain. The main use of acupuncture is to treat pain. The reason it works so well is that during an acupuncture session, hormones known as *endorphins* are released in the brain; these have similar characteristics to painkilling drugs such as morphine. Neurotransmitters such as serotonin and adenosine are also released, which can help tackle pain and alleviate mood, and help with the immune system.

Acupuncture works by improving blood circulation in the body, and helps balance the autonomic nervous system (ANS). The ANS is a control system responsible for regulating bodily functions such as the heartbeat, digestion, urination, respiratory rate and sexual arousal (see Chapter 8 on the brain and mindfulness for more about this).

Qi circulation is a network of invisible pathways called *channels* or *meridians*, which span the whole body, each connected to one of the 12 main organs: the liver, the kidneys, the heart, the lungs, the stomach, the intestines, the brain, the reproductive organs, the thyroid, the gall bladder, the skin and the bones.

There are 365 points on the body where acupuncture needles can be inserted. Qi flows through each of the 12 main channels every 24 hours; it takes 2 hours to pass through each. This is known as the *Chinese clock*. Qi circulation is compared to circulation of blood. Chinese medicine describes every interaction in the universe as *yin and yang energy*. Yin and yang energies are opposites, interdependent; they consume and transform, and both must be balanced in order for a person to feel healthy and happy.

» Characteristics of yin and yang

Yin: water, sun, cold, wet, dark, passive, slow, soft, feminine, descending, destructive, darkness and death. Yin organs are solid, and include the liver, spleen, heart, lungs and kidneys.

Yang: fire, moon, hot, dry, light, active, rapid, hard, masculine, rising, generative, light and creation of life. Yang organs are hollow, and include the gall bladder, small and large intestines, stomach and bladder.

The Chinese believe that when an illness occurs, the yin and yang energies are out of balance and the use of acupuncture can help restore the balance.

Disease, injury or trauma can interrupt the natural flow of qi within the meridians/organs. This can cause:
- stagnation of qi, causing pain and inflammation
- deficiency of qi, causing weakness and exhaustion.

An acupuncture treatment can include *acupressure* – the practitioner uses their hands on the same meridian points and applies pressure. This relieves muscle tightness, stimulates qi flow and restores balance, moxibustion (burning of dried mugwort close to the skin) and electroacupuncture (use of low voltage passing between two needle points). To train as a practitioner or find a registered acupuncturist in the UK, visit the British Medical Acupuncture Society's website at www.medical-acupuncture.co.uk.

» Reiki

Reiki healing is a unique healing system whereby the practitioner transmits a spiritually guided 'life force' through their hands during the treatment. It is a non-invasive method to help balance and restore energetic aura. It can be used to treat humans and animals. A Japanese doctor developed reiki by using his body as a channel of energy which was transmitted to his hands. The word 'reiki' comes from the Japanese terms *rei* (universal consciousness) and *qi* (life energy). The human body is surrounded by a field of energy which, as we now know, is the energetic aura. Reiki works by combining with the energetic aura of the patient, and brings comfort to them by giving them a feeling of calm and relaxation.

Reiki is a gentle but extremely powerful way of healing through the human energy field (HEF), which can treat physical, emotional and spiritual imbalances. As reiki is a conscious life energy, it is able to detect and reach the root cause of the problem. An advanced practitioner can perform aura cleansing, where the healing energy begins at the top of the head and travels down the body with hand positions slightly above the body: this is why it is a non-invasive method. A reiki practitioner helps to make more life energy available to the receiver, so self-healing of the body and the mind can be activated. This will help the receiver feel healthier.

It must be noted and emphasised that reiki is not a replacement for any medical treatment; please refer to a doctor for medical advice for any condition or illness you have.

Benefits of reiki:
- stress relief
- deep relaxation
- pain relief
- detoxification
- strengthening of natural self-healing pores
- mind–body connection
- connected to all sources of life energy
- releasing any blockages
- peace
- balance
- rejuvenation
- spiritual connection.

To get the full benefit of reiki, two to six treatments are recommended – the number of treatments needed will be assessed by the practitioner.

To enhance a reiki treatment, aura cleansing and chakra balancing can be incorporated into the session.

» Animal reiki

I am an animal reiki practitioner. In this practice, I use the same technique and attuned symbols as reiki during my sessions. I have done distant healing and face-to-face healing.

A case study I did was with a dog called Sphinx belonging to a close friend of mine. The dog was taken out for a walk by her partner. On the way back, he crossed the road and the dog was on the lead behind him; the dog was hit by a car from behind. The dog was badly injured, and my friend was worried that he might die. She sent me a picture of the dog, and I started the distant healing straight away.

Figure 2.2: Reiki symbols

The healing works through me connecting my energy with the energy of the dog and using the attuned symbol for distance healing (see Figure 2.2). I was sending the healing from my heart chakra, as well as having the intention to heal the dog. The dog was seen by a vet, who explained to my friend that he might die. My friend took the dog back home, feeling very sad. I sent her and the dog distant reiki healing and arranged to come and visit the dog. When I visited, the dog was quite lethargic; he looked extremely sad and actually started to growl – since the accident, he had not been happy being near humans, especially her partner. As the evening progressed and we had our dinner, the dog got closer and closer to me, and eventually sat by my feet for the rest of the evening. I was giving him healing the whole time I was there. After 24 hours, the dog stopped limping and was feeling much better. Against all the odds, the reiki healing I gave him really helped his recovery, and he is still healthy to this day.

Below is the testimonial written by my friend regarding the healing I gave her dog:

> *My dog (Sphinx) was hit by a car and had incurred a hip trauma and internal bleeding which had left him with a swaying hip when he walked. He was also in a lot of pain and was not able to walk for more than 15 minutes without panting in pain due to his internal injuries. Sphinx had one healing session with Maria, and he totally surrendered to the healing energies during the session with Maria and seemed to really love the healing vibes. Twenty-four hours later, Sphinx not only had a spring back in his step, but he managed to walk for an hour without showing any sign of pain and seemed to be back to his old happy self again. I highly recommend Maria for animal healing as she has a natural gift with animals, and animals really do soak up all that love and healing energy!*
>
> Sam

I end this chapter in Glastonbury, where I came to enjoy the therapeutic energies of the white spring and red spring. I have gone through a massive transformation since the death of my father, and I can truly say it was the hardest thing I have ever been through. Losing a parent makes an individual evaluate their life and be grateful for everything they have. I have seen so many different layers of my energetic aura and been through a spectrum of emotions: sad, guilty, depressed, anxious, abandoned. It is so important to look after your health and well-being, especially through difficult times. I hope this book will help you and give you the skills you need to achieve this.

Positive affirmation

I am energy! I am balanced, vitalised, healthy and happy! I am connected to the air, the water, the sun, the earth! I am life! I am love!

Activity 1: *Connecting to your energetic aura*
- Find a quiet place.
- Think of your favourite colour.
- Find an object of that colour, or you could just imagine that colour.
- Lie down and close your eyes.
- Concentrate on your breathing, then imagine you are this colour. What does it feel like? Does it have a smell? What does it look like?
- Spend about 10 minutes visualising this colour.
- Open your eyes and write a short paragraph about this colour in your journal. This is the general colour of your energetic aura; refer to Table 2.1 for the characteristics of that colour, and see if this resonates with you.

Activity 2: *Connecting to another energetic aura*
- Find a quiet place.
- Decide on what energetic aura you would like to connect to, be it that of a pet, the moon, the sun, nature or a family member.
- Hold an object or an image of what you want to connect with.
- Sit upright in a chair and ensure your feet are firmly on the ground.
- Close your eyes and connect to the energetic aura of the object or image.
- Concentrate on your breathing.
- Imagine yourself in a rainbow bubble, and imagine that you are floating up in the sky with the energy you are wishing to connect with.
- The bubble lands in a beautiful forest with waterfalls, lots of trees, birds singing, a nice breeze and the sun beaming on your face.
- Spend 10 minutes connecting with the energetic aura.
- When you feel ready, open your eyes and journal your experience.

Reading

Boland Y. *Moonology: Working with the Magic of Lunar Cycles*. Hay House, 2016.

Eason C. *A Little Bit of Auras: An Introduction to Energy Fields*. Sterling Publishing, 2018.

Hicks A. *The Acupuncture Handbook: How Acupuncture Works and How It Can Help You*. Piatkus Books, 2005.

Kaufmann M. *Understanding reiki*. First Stone Publishing, 2004.

Mercier P. *The Chakra Bible: The Definitive Guide to Working with Chakras.* Octopus Publishing Group, 2009.

Norris S. *Secrets of Colour Healing.* Ivy Press, 2018.

Useful web pages

Wikipedia: Moon – en.wikipedia.org/wiki/Moon

Wikipedia: Sun – en.wikipedia.org/wiki/Sun

Chalice Well Trust – www.chalicewell.org.uk

CHAPTER 3

The human body

The aim of this chapter is not to overwhelm you with lots of technical information, but to give you a greater understanding of how the human body works. It covers two themes: the structure and function of the human body, and what is needed for optimal health.

The human body is a very complicated and magnificent machine, consisting of billions of tiny circular balls of energy called *cells*. A cell is the basic functional unit of the body, containing three main organelles: a nucleus (the control centre of the cell – genetic make-up), the cytoplasm (where chemical reactions take place) and the cell membrane (its protective barrier). The reactions that occur in the cytoplasm involve energy release, which is needed to support life and create structures. Each cell contains coded instructions to control cell activities.

Cells can be created and recreated by a process called *cell division* (*mitosis*). For example, when you cut your finger and it starts to bleed, cells in the blood called *platelets* will start to clot the blood, which will cover the cut; after a few days, a scar can be seen, made of newly created skin cells. It is possible to visualise the ultrastructure of a cell by using an electron microscope, making the subcellular organelles visible. Each organelle has a unique structure and function within that cell. To understand the different structures and functions of each organelle in an individual cell, please refer to Figure 3.1 and Table 3.1.

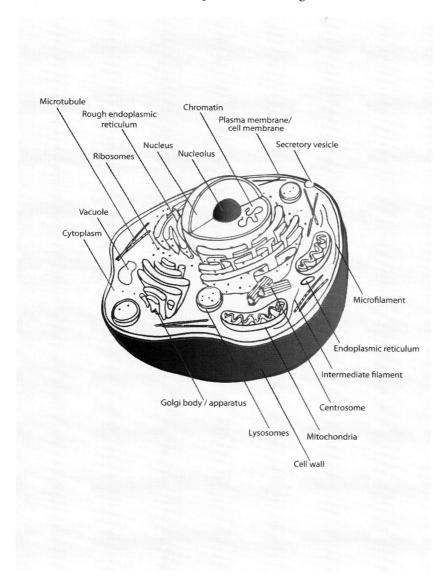

Figure 3.1: The cell

Table 3.1: Main cell organelles, with their structures and functions

Organelle	Structure	Function(s)
Cell membrane	Double phospholipid layer with extrinsic and intrinsic proteins	Protects the cell
		Separates the cell from the outside environment
		Regulates the passage of molecules into and out of the cell
Nucleus	Round organelle with a nucleolus and nuclear pores	Controls all cellular activity in the cell
		Stores DNA* and controls cell metabolism
		mRNA** leaves nuclear pore for protein synthesis
Cytoplasm	Semi-liquid medium	Dissolves many ions
		Many chemical reactions take place
Ribosomes	Small circular organelles attached to the rough endoplasmic reticulum	Protein manufacturing
		Site where mRNA meets tRNA*** for amino acid bonding
Endoplasmic reticulum (ER)	Two types: rough (with ribosomes) and smooth (without ribosomes)	Smooth ER aids synthesis and transport of lipids
	Series of folded internal membranes	Rough ER produces proteins
Golgi apparatus	Series of flattened sacs similar to a packaging system, giving chemicals a membrane to form into a vesicle	Aids the production and secretion of many proteins, carbohydrates and glycoproteins
Lysosomes	Specialised vesicles containing digestion enzymes	Main function is digestion
		Enzymes break down proteins and lipids

Centrioles	Two short cylinders containing microtubules	Aid cell division
Mitochondrion	Double membrane enclosed in a semi-fluid matrix	Carries out aerobic respiration to release energy into the cell known as *ATP*
Microfilaments	Solid rods made up of actin	Help in structural support of the cytoskeleton
Microtubules	Straight hollow cylinders in the cytoplasm	Many functions: transport and structural

* DNA – Deoxyribonucleic acid is a double-stranded nuclear chain found in the nucleus. DNA makes a copy of itself from the original DNA strand in a process called *DNA replication*.

** mRNA – Messenger ribonucleic acid is a single-stranded nuclear chain made in the nucleus. It copies unzipped strands of DNA, which it takes to the cytoplasm in a process called *transcription*.

*** tRNA – A short length of RNA needed to bind to mRNA to code for one amino acid. Amino acids are the building blocks that make proteins through a process called *protein synthesis*.

What is DNA?

Deoxyribonucleic acid (DNA) is the component that stores all the genetic information about the origin and specific function of the cell. The structure of DNA is known as a *double alpha-helix* consisting of two chains of nucleotides interlinked together and bound by hydrogen bonding. Each nucleotide contains a sugar, a phosphate group and one of four bases: adenine (A), guanine (G), cysteine (C) and thymine (T). These bases are complementary: A pairs with T and C pairs with G, so one strand will have base A and the other will have base T.

It is possible to extract DNA from a sample of pathological tissue to look at specific genes, mutations and chromosomes in that specimen. A mutation is a change to a specific part of a gene caused by a deletion, an insertion or a rearrangement in the genetic code. This alteration can play a part in normal processes in the body such as evolution, and in abnormal ones such as cancer. A molecular genetic technique called *DNA extraction*, involving extracting DNA from blood, human tissue and other biological fluids, is used to identify mutated genes specific to a type of cancer and inherited genes that may indicate whether someone is likely to get that specific cancer. For example, if someone's mother has BRCA1 and BRCA2 genetic mutation breast cancer, they will have a higher-than-average risk of developing this disease. DNA is unique to each person, and is the foundation of who they are in every way: genetically, biologically and spiritually.

What is a tissue?

A tissue is the organisational level containing an extracellular matrix woven in between the cells and the organ. Each tissue is specific to the type of organ. *Histology* is the study of human and animal tissue, while *histopathology* is the study of disease in tissue, such as cancer. I have hands-on experience in this field – I have worked in a histopathology laboratory for most of my scientific career.

A histopathological assessment involves removing the tissue from the animal or human. This can be done through a biopsy (needle) or an excision (a large part of the tissue is removed, or the whole organ). After removal, the specimen is placed in a fixative called *formalin*. The fixative is used to preserve the tissue and keep it as intact as possible. The tissue then goes through a sequence of clearing (xylene), dehydration (alcohol) and fixation; then it is placed in a small block cassette and embedded in paraffin wax. Sections are cut from the block using a machine called a *microtome* and placed on glass slides. The sections are dried in a 37°C oven overnight. The next step is to stain the section with a specific dye called *haematoxylin and eosin stain* (*H and E*) to identify the nucleus, cell membrane and cytoplasm of each cell in the tissue. A trained histologist, such as myself, will be able to identify the normal and diseased cells in the section and make a diagnosis using a light microscope.

Another technique that can be used to assess tissue is *frozen sectioning*. A biopsy is taken at surgery while the patient is still on the operating table, so the pathologist can make an instant diagnosis; this is usually done if there is a suspected cancer. Once the biopsy has been removed, it is sent to the laboratory, where a section of tissue is taken using a machine called a *cryostat*. The biopsy needs to be frozen in liquid nitrogen, and is then embedded in OCT (optimal cutting temperature – a medium used to mount the specimen). The H and E is looked at by an experienced scientist to confirm tumour, and sent to a pathologist to confirm diagnosis. The outcome of this is that the surgeon may remove more tissue than intended, or the whole area of tumour/organ.

What are the different types of tissue in the human body?

There are four types of tissue found in the body:
- **Epithelial** – Covering all body surfaces and lining body cavities and hollow organs.
- **Connective** – Binding and connecting the organs.
- **Muscular** – Moving body parts. Movement can be voluntary (under conscious control, such as deciding you will read a book on anatomy) or involuntary (not under conscious control, such as the pupil in the eye contracting in bright light).
- **Nervous** – Sending and receiving electrochemical signals that provide the body with information.

Each organ in the body is made up of specific cells and tissues; it has a specific function and controls individual body systems. For example, the function of the heart is to pump oxygenated and deoxygenated blood around the body; this function controls the respiratory and cardiovascular systems. Refer to Table 3.2 and Figure 3.2 for the main organs in the body. Table 3.2 includes a short description of the structure and function of each organ.

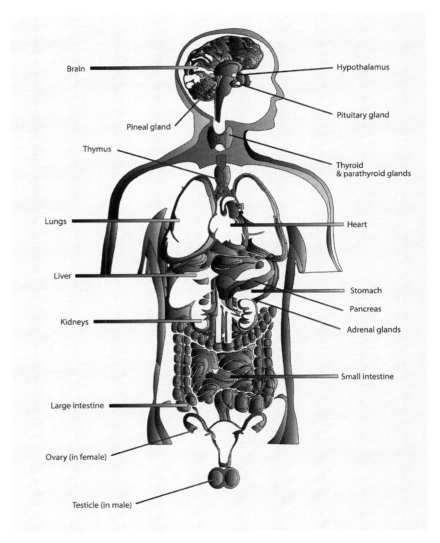

Figure 3.2: Structure and location of each organ in the body, including endocrine glands

Table 3.2: Main organs in the body, with their structures and functions

Organ	Structure	Function(s)
Brain	Walnut-shaped with billions of neurons and neuronal tissue	Controls all the neuronal activity in central nervous system (CNS) and peripheral nervous system (PNS)
Thymus	Two-lobed structure in the upper chest cavity	Immunity
Lungs	Two large airbags with bronchi	Gaseous exchange
Heart	Muscle consists of four chambers (two atrium, two ventricles)	Pumps blood around the body
Liver	Multi-lobed structure sitting under the diaphragm. The largest organ	Detoxification of the blood, bile production, glycogen production and removal of old red blood cells. Ability to regenerate
Pancreas	Long, flat organ located near the duodenum	Secretes hormones (glucagon and insulin) and digestive enzymes such as pancreatic amylase to continue starch digestion
Stomach	Kidney bean structure with tube connected to the oesophagus	Digestion
Kidneys	Two oval-shaped structures connected to the dorsal aorta and renal artery and vein	Excretion. Production of urine
Small intestine	Consists of three tubular structures: duodenum, jejunum and ileum	Portal of absorption of nutrients in the bloodstream. Breaking down of large molecules to smaller molecules

Large intestine	Long, spiral tube consisting of the colon and rectum	Final stage of digestion: recovery of water and electrolytes, formation and storage of faeces
Ovary (female)	Two egg-shaped structures	Site of reproduction Releases oestrogen and progesterone hormones
Testes (male)	Two ball-shaped structures	Site of reproduction Releases testosterone

The endocrine system and hormones

The *endocrine system* in the human body is a complex mechanism; it is the system that secretes substances called *hormones*.

A hormone is a chemical substance that is needed for specific processes in the body to function properly. Each hormone is produced by a specialised gland. Once the hormone has been secreted, it travels in the blood to reach the target organ; a response in the organ is triggered by the activation of a particular enzyme – a type of protein that is specific to a certain reaction. Table 3.3 gives a description of each endocrine gland (structure, location and function(s)), including which specialised hormone(s) it secretes.

Table 3.3: Structure, location and functions of each endocrine gland

Endocrine gland	Structure and location	Function(s)
Hypothalamus	Small cone-shaped body that projects downwards from the brain	Link between the nervous system and endocrine system via pituitary Controls and regulates body temperature, sleep and cardiac cycle

Pineal	Small glandular body in the brain	Secretes melatonin required for the sleep cycle
Pituitary	Pea-sized organ in the brain, with two parts: anterior and posterior	Anterior produces seven hormones: growth, prolactin, TSH (thyroid-stimulating hormone), ACTH (adrenocorticotrophic hormone), FSH (follicle-stimulating hormone), LH (luteinising hormone), MSH (melanocyte-stimulating hormone) and ADH (antidiuretic hormone)
		Posterior stores antidiuretic hormone (kidney function) and oxytocin (produced while giving birth)
Thyroid	Two lobes located either side of the larynx	Controls metabolism
		Produces thyroxine
Parathyroid	Size of two grains of rice; located in the thyroid gland	Secretes parahormone
		Regulates calcium metabolism
Adrenal	Located near the cranial poles of the kidneys, with two parts: adrenal cortex, adrenal medulla	Cortex (under the control of ACTH (adrenocorticotrophic hormone) secreted) produces three hormone groups: glucocorticoids (cause increase in blood sugar), mineralocorticoids (regulate salt ions in the kidney: sodium, potassium and hydrogen), sex hormones
		Medulla (under the control of the sympathetic nervous system) releases adrenaline

The endocrine system will be discussed further, together with how it works alongside the chakra system, in the next chapter. It is not the only system in the body; the other systems are summarised in Table 3.4.

Table 3.4: Body systems, with their organs, body parts and functions

Body system	Organs/body parts involved	Function(s)
Skeletal	Bones, joints, muscle	Internal support, protection and body movement
Digestive	Mouth, oesophagus, stomach, intestines, pancreas, rectum	Breaks down food into nutrient molecules; these enter the cells via the blood
Cardiovascular	Heart, lungs, blood	Transports nutrients and oxygen to cells and removes any waste products from the cells
Immune/lymphatic	Immune cells, white blood cells, lymph nodes, lymphocytes	Body's defence system against bacteria, viruses and pathogens, drainage of fluid
		Protects body from disease by purifying the white blood cells (lymphocytes) that produce antibodies and combat disease
Respiratory	Lungs, blood, nose	Takes in oxygen and removes carbon dioxide
		Helps maintain PH in the blood
Urinary	Kidneys, blood	Gets rid of unwanted nitrogenous waste and helps in regulating fluid levels in the body

Nervous	Brain cells, CNS, PNS, hormones	Regulates all the other systems
		Instructs the body to do everyday tasks, memory, cognitive function
Endocrine	Brain, hormones, hypothalamus, glands: pituitary, thyroid and parathyroid, adrenal, pineal	Eight major glands that secrete hormones into the blood. These hormones regulate bodily processes in the body such as metabolism
Excretory	Rectum, anus	Removal of waste such as urine and faeces
Reproductive	Testes (male), ovaries (female)	Produces offspring – production of eggs or sperm

I have given an overview of the structures and systems in the human body, and how these systems work together. There are many processes and requirements that are needed for the body to function in a balanced, healthy and optimal way. This will be explained in the next section of this chapter, along with answers to specific questions.

What is the process that regulates body systems?

Homeostasis is the process that regulates the body's natural environment, such as body temperature (which must always be in the range from 36.1°C to 37.2°C), blood PH and blood sugar levels. This is achieved by a negative feedback process. The process involves a sensor that detects changes in the internal environment, and a regulatory centre that activates the effector; this reverses the changes and brings the condition back to normal again.

An example of this is body temperature; this needs to be maintained in the 36.1°C to 37.2°C range because all enzymes in the body only work optimally at this temperature (enzymes are required for all chemical reactions in the body). Temperature is kept balanced by the hypothalamus in the brain, which acts like a thermostat for the body. When the body temperature changes because of external conditions such as extreme heat or extreme cold, the hypothalamus works hard to restore balance in the body by initiating a series of specific processes – for example, when it is very hot, our body produces sweat, which helps to cool us down.

What nutrition do your cells need to be healthy and function well?

All the cells in the body need macronutrients – carbohydrates, proteins and fats – which are necessary to supply energy and are building blocks to maintain cellular activities.

They also need micronutrients – vitamins and minerals – which are needed in the right quantities to eliminate illness. For example, vitamin C deficiency can lead to scurvy.

» Daily requirements

- **Energy**: 2,000 (female) / 2,500 (male) calories (less if you need to lose weight: approximately 1,500 for women, 2,000 for men; remember that to maintain a steady weight, you must consume the same amount that you burn).
- **Carbohydrates include**: rice, pasta, vegetables, fruit – 260g.
- **Proteins include**: eggs, dairy, cheese, chicken, red meat, fish, beans, pulses – 50g.
- **Fats include**: olive oil, butter, avocado – 70g.

How does our body make and store energy?

Aerobic respiration is the process in which the body makes energy. It involves the breakdown of glucose (a carbohydrate) to carbon dioxide (CO_2) and water (H_2O). Aerobic respiration consists of three processes: glycolysis, kerb cycle and oxidative phosphorylation. These processes are vital in making ATP (energy molecules).

Glycolysis takes place in the cytoplasm, the *kreb cycle* takes place in the matrix of the mitochondrion, and *oxidative phosphorylation* takes place in the cristae of the mitochondrion. During these processes, a series of chemical reactions occur, where the product of the previous reaction becomes the substrate for the next one. Every single reaction requires a specific enzyme, which speeds up the rate of reaction to produce the final product. All three pathways work at the same time. During the whole process of cellular respiration, hydrogen is removed to produce ATP, water and carbon dioxide.

ATP is a complex type of molecule that provides energy to cells and aids many cellular processes including muscle contraction and DNA replication. The key organelle in this process is the mitochondrion (see Table 3.1).

What are senses and how do they work?

Sensory integration is the way that the brain organises sensations from the body and its environment. This makes it possible to use the body in a way that creates adaptive responses to the environment. Sensory processing involves sensory receptors in the brain recognising certain stimuli, interpreting and organising the input at any given time.

Please see Table 3.5 on the next page for a detailed description of each sense, along with its spiritual counterpart and how the sense can be activated.

Table 3.5: Activating your senses

Sense	Spiritual sense	Receptor	Stimulus
Vision/sight	Clairvoyance	Rod and cone cells in retina of eye	Light rays from the electromagnetic spectrum
Smell	Clairalience	Olfactory cells	Chemicals, odour
Taste/gustation	Clairgustance	Taste Cells	Chemicals
Hearing	Clairaudience	Hair cells in spiral organ	Sound waves from the electromagnetic spectrum
Touch	Clairtangency	Somatosensory	Activation of neuronal receptors in the skin
Rational equilibrium	Clairaudience	Hair cells in semi-circular canals	Motion
Gravitational equilibrium	Clairaudience	Hair cells in vestibule	Gravity

Organ	Body system	Function	Activation
Eye	Visual	• Capacity of eyes to focus and detect images of visible light • Generate electrical nerve impulses for varying colours and brightness (day/night)	• Compare similar objects • Look at matching objects • Spend time looking at all the colours of a rainbow and a prism • Separate overlapping images in a busy picture • Making sense of partially visible objects • Look at black-and-white images and compare colour and contrast
Olfactory epithelium	Olfactory	• Hundreds of olfactory receptors in the nose detect all sorts of odours • Increases awareness of environment • Brings joy when smelling something you like	• Walk through a sensory garden • Use smell boxes for food, plants and flowers • Use essential oils to make sprays or meditation oils with different scents
Taste buds	Gustation	• Capacity to detect taste of food, chemicals and poisons • Taste buds: sweet, bitter, sour, salty	• Try different types of food • Use spices and herbs in your cooking • Eat a variation of sweet and sour dishes • Try food from different cultures

Organ	Body system	Function	Activation
Ear	Auditory	• Ability to perceive sound by detecting vibrations • Sound can be differentiated according to location, type, memory	• Use headphones to listen to all types of music and sounds • Listen to the flow of water • Go out into nature, connect with all the sounds there and write them down • Make sounds using an echo box
Skin	Somato-sensory	• Helps to feel different sensations by touch to give us different emotions • Differentiates between soft, hard, hot, cold, high pressure, low pressure	• Stroke your pet; how does this make you feel? • Touch ten different objects and write down how they feel • Roll around in grass, snow, carpet and sand • Swim in the sea, river, pool
Ear	Vestibular	Coordinates movement with balance by spatial orientation, head balance and motion	• Active movements such as jumping, spinning, headstand • Stand on one foot and keep your balance
Ear	Vestibular	Coordinates movement with balance by spatial orientation, head balance, motion and gravity	Test your balance at various heights: for example, lower ground, uphill or even higher on a mountain top

What is needed for a healthy and fit body?

- **Nutrition** – Having a well-balanced diet with fish, meat, vegetables, fruit, nuts, pulses and grains. Avoid sugary foods and processed food (overconsumption of these can lead to obesity, hypertension, diabetes and even cancer).
- **Exercise** – Exercise at least 1 hour a day. This doesn't have to be vigorous exercise – walking and swimming are just as effective as aerobics and running. Exercise will help with blood circulation and will give your cells more energy. It improves muscular strength, muscular endurance and flexibility, and can protect against certain diseases such as cancer. Bone density can be improved by physical training, such as cardio combined with resistance training (using weights). Exercise can also help with depression by elevating your mood. When you exercise, your brain releases endorphins; these make you feel happy and good about yourself. As a result, your energy level increases and you will feel more active.
- **Sleep** – Getting enough sleep is very important. Too much or too little can affect your natural body clock. An average of 8 hours a night is sufficient for most people.
- **Rest** – Rest is different from sleep; it is important to go out and connect to nature and the four elements – water, air, earth, fire – as much as you can. Try to meditate for 10 minutes a day; do this at the same time every day, and include it in your daily schedule. Go on holiday a couple of times a year, to get away from the hustle and bustle of daily life. Spend time with friends and family to feel loved, happy and supported.

What happens when things go wrong?

The body can go wrong, and illnesses can occur, due to genetic, environmental and personal choices.

Let me give you an example of each:
- **Genetic** – These arise from a single mutation in a gene, and are passed on from your parents or grandparents – for example, BRCA1 and BRCA2 breast cancer.
- **Environmental** – Stressful situations account for many illnesses in the human body, such as anxiety and depression.
- **Personal choice** – If someone decides to smoke, this can increase their chance of getting lung cancer.

» Common conditions occurring in the body

- **Cancer** – The most common disease. It occurs when there is a genetic mutation in a cell; the cell division process becomes extremely rapid, causing the cells to divide in an uncontrollable way, changing the normal shape and structure of the cell. Normal cells turn into abnormal ones called *tumour cells*. After a short time, a mass is formed called a tumour. A tumour can develop in any organ of the body, and tumour cells can be found in the blood. Cancer of the blood comes in different forms: lymphoma and leukaemia.

  It is still unknown how cancer occurs, though there are many contributing factors, including nutrition, genetic make-up, exposure to toxins, smoking and overconsumption of alcohol. This is one of the main diseases that can arise when cells go wrong, and can interrupt normal functioning of the body, making the individual very unwell. Treatment can be given, including chemotherapy, radiotherapy and hormone positive treatment. These treatments can help the individual, but the side effects can be severe.
- **Obesity** – The most common medical condition in the western world. Overconsumption of fatty and sugary foods leads to layers

of fat around the body, resulting in problems with vital organs and body systems. Obesity is defined by body mass index (BMI) and can be calculated by this formula:

Body mass index = body weight (kg) ÷ height (m^2)

If your calculated BMI is over 30, it indicates that you are obese; you must try your best to reduce your weight by healthy eating and exercise, otherwise you will be at an extremely high risk of developing type 2 diabetes and heart disease.

- **Cardiovascular diseases** – These are diseases of the circulatory system, including the heart and blood vessels. The reason behind these diseases is a gradual deterioration of the heart and blood vessels; fat deposits occur in the arteries, which can accumulate over time and prevent the blood from going around the body efficiently. This could lead to heart disease, heart attacks or difficulty breathing because oxygen supply is reduced.
- **Hypertension** – High blood pressure is very dangerous and can cause heart failure or a stroke. It is often called the 'silent killer', because there are no symptoms. High blood pressure is linked to excessive alcohol intake, smoking, obesity, high salt in the diet, and genetic factors (for example, if your parents have hypertension, you may have a genetic disposition for the disease).

It is extremely important to look after your body and keep it healthy – just as it is to keep your energetic aura balanced and healthy, along with your mental health (this will be discussed in more detail in Chapter 8: 'The brain and mindfulness'). All three – mind, body and soul – must be in synchronisation and balanced, and must work together for optimum health. In the next chapter, we will explore the chakra system.

Activity 1: *Well-being diary*
Use this template to keep a well-being diary for a week, recording food, exercise, sleep and rest, to make sure you are in the ideal range for optimal health.

Item	Day:
Carbohydrates (type, amount)	
Proteins (type, amount)	
Fats (type, amount)	
Water (amount)	
Sleep (hours)	
Rest (minutes)	
Exercise (type, intensity, length)	

Activity 2: *Visualisation*

Visualisation – Make yourself comfortable, either sitting upright or lying down, and close your eyes. What I want you to do is to connect to all the cells in your body. Every cell is precious and essential to your well-being; connect with and send love to your cells, and imagine that they are all functioning and vibrating correctly. Imagine every tissue, organ and system in your body is healthy and vibrant. Your body is your temple!

Affirmation – I am healthy, rested and relaxed! I am supported and protected!

Reading

Clegg CJ, Mackean DG. *Advanced Biology: Principles and Applications*. John Murray Publishers, 1994.

Fosbery R. *Human Health and Disease*. Cambridge University Press, 1997.

Mader SS. *Human Biology: 7th edition*. McGraw-Hill Higher Education, 2002.

Parker J, Honeysett I. *Revise AS and A2 Biology: Complete Study and Revision Guide*. Letts Educational Ltd, 2008.

Wingate P, Gifford C, Treays R. *Essential Science*. Usborne Publishing, 1992.

CHAPTER 4

The chakras

This chapter begins on my summer holiday in Fuerteventura. I am sitting on my balcony enjoying all the energetic auras around me: the sounds of the waves crashing onto the sand, the ocean breeze and the amazing vibrant aura of the harvest full moon. This place is amazing: peaceful, tranquil and pure relaxation. I feel balanced, rested and calm!

In this chapter, I will take you on a journey through the chakras. You will learn about each chakra in detail, and how it works with a specific gland in the endocrine system to reach optimal equilibrium.

What is a chakra?

A chakra is a vortex of spinning energy found in the subtle body. It is created when two or more lines of energy cross. The word 'chakra' comes from the Sanskrit word meaning wheel. The chakras govern the energy fields of the cells, organs and bodily systems, in particular the endocrine system. The more you connect with and understand your own chakra system, the more connected you will feel with the mind, body and soul in a complete cycle of vibrant energy and good health.

The lotus

Different types of lotus flower represent the various chakras in the body, as they blossom within the auric field. The lotus flower grows in muddy waters on the surface of the lake and blossoms in the light (sun = rebirth), with long stems that grow deep down in the soil at the bottom of the lake. It symbolises growth, determination, cleanliness, wholesomeness, fortune and commitment. The lotus blossoms into a magnificent flower, even in a difficult environment.

Many religions ascribe a specific symbolic meaning to the lotus flower. In Buddhism, for example, it symbolises the purity of the body and mind, attachment, spiritual enlightenment and ascension. In Christianity, it represents purity and fertility; while in Hinduism, it symbolises the connection to the Hindu gods and goddesses. There are many different colours of the lotus, each with a symbolic meaning: for example, white is associated with purity, yellow with enlightenment, red with love and compassion, and blue with ascension and rebirth.

The chakras

In Table 4.1, I describe the chakras as they appear on my logo (Figure 4.1), starting from the bottom (chakra 1 – earth star) ascending to the top (chakra 13 – stellar gateway).

Chakras marked with '*' are known as the *newly emerging* chakras; these are already activated in some people on their path of ascension leading to enlightenment. The three celestial chakras above the crown extend out of the physical body chakras, and are closer to the last layer of your auric field. In this chapter, you will learn how to activate these newly emerging chakras: the earth star, the hara, the higher heart, the causal, the soul star and the stellar gateway.

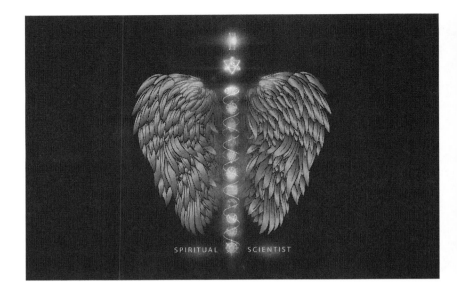

Figure 4.1: Spiritual Scientist logo. The double helix of DNA runs down the middle in between the chakras

Table 4.1: The 13 chakras, their colour frequencies, energy types, corresponding body systems and endocrine glands, and locations in the body

Chakra	Colour frequency	Energy type	Body system/ endocrine gland	Location
*Earth star	Magenta (black before activated)	Earth	The brain (mind)	Below the base of the feet; extends out forming a circle of energy
Root	Red	Grounding	Reproductive system (ovaries in females, testes in males)	Base of the spine

Chakra	Colour frequency	Energy type	Body system/ endocrine gland	Location
Sacral	Orange	Creativity	Nervous system (adrenal gland)	Below the navel
*Hara	Mix: orange/ yellow	Qi energy	Digestive system (kidney)	Above the navel
Solar plexus	Yellow	Power	Digestive system (pancreas)	The navel
Heart	Green	Healing	Immune system (thymus)	Centre of the heart
*Higher heart	Pink	Unconditional love	Immune system (thymus)	Just above the centre of the heart
Throat	Blue	Communication	The endocrine system (thyroid and parathyroid)	Centre of the throat
Third eye	Purple	Intuition	The brain (pituitary)	In the middle of the brow
Crown	Violet, gold, white	Connection to spirit	The brain (pituitary and pineal)	At the top of the head
*Causal	Turquoise	Higher consciousness	The brain (mind)	10cm above the head
*Soul star	Pink	Enlightenment and ascension	The brain (mind)	15cm above the head
*Stellar gateway	Sparkly white light	Universal consciousness – masculine and feminine combined	The brain (mind)	30cm above the head

Ascension

This is the process of evolution of the soul consciousness and the connection with higher dimensions. It is a process of spiritual awakening. During this process you will go to a higher state of consciousness and your vibrational frequency will become much lighter.

This requires disregarding the ego and becoming more in touch with the vibration of love and light. It involves you being much more aware of the environment, the seasons, the moon, the sun, the stars, the planets; and your connection to other light beings in other realms, such as the angels and archangels. This process will not happen overnight – it could take months or even years. You may experience certain symptoms during your journey that may cause you discomfort, such as:
- heightened sensitivities to the senses: sight, hearing, smell, taste and touch
- feeling more empathy towards others and taking on their sadness and upset
- aches and pains
- flu-like symptoms
- new intolerances to food, drink and certain fabrics.

To overcome these, make sure you stay balanced, protect your energy and connect to nature as much as possible.

The earth star chakra

This chakra, when activated, works alongside the root chakra to help you feel grounded and connected to Mother Earth on this physical plane. This chakra also helps you to connect to animals, nature spirits such as the trees and the flowers; and helps you to show more love

and compassion to all of your environment. You will start to protect the environment in any way possible, such as:
- helping in beach clean-ups
- using less or no plastic
- having meat-free or animal-product-free days
- respecting the animals that you do eat, and giving gratitude for the energy they provide for you
- having a more basic life
- releasing the negative ego.

The negative ego is one of fear, jealousy, anger, resentment and pain. Higher-frequency energies for the newly emerging chakras descend downwards to this chakra and cause it to expand outwards. The earth star shines a beautiful magenta colour when vibrating! This enhanced auric field, in combination with the crown and the three other major chakras above the head – causal, soul star and stellar gateway – will help us ascend into the fourth and fifth dimensions, and bring back memories of our lives among the stars and the beginning of time. This is in preparation for light body ascension and our past lives in ancient civilisations such as Lemuria and Atlantis.

Activity: *Activating your earth star chakra*
- Go for a walk barefoot in a forest filled with trees, wild flowers and animals. Make sure your feet are firming flat on the ground, so that the earth energy can connect with your energy. This will ground you deeply into your natural environment, and will charge the ground you tread with great light and energy.
- Pay attention to the sound, the smells and the beauty of the natural environment.
- Find a nice quiet spot and sit down with your feet still connecting to the earth energies.

- Close your eyes, and pay attention to your breathing, and really feel into all the energies surrounding you.
- Start your meditation and stay there for at least 30 minutes.
- During your meditation, you may be taken to the same land in a previous life or time. It is as if you are entering a time machine, and you may go back to the land of Lemuria, Atlantis or even Avalon.
- Pay attention to colours, images or people you may see.
- You will need to do this exercise a few times before you start to get visualisations, and before this chakra is fully activated. To enhance the activation, work with clear quartz crystal during the meditation; this is the main healing crystal, and will amplify your connection to the earth and surrounding energies.

» **Affirmation**

I am open to all the energies of nature! I honour Mother Earth and all of her inhabitants!

(My story of activating this chakra and my connection to the ancient land of Atlantis will be described at the end of this chapter.)

The root chakra

This chakra establishes our connection to Mother Earth and all the elements. It is activated in the early stages of life, beginning in the womb with the connection we have with our mother. We rely on our mother for survival during pregnancy – she provides us with all the nutrients to grow via the umbilical cord. Once we are born, this connection is still strong and we rely on our mother and father for food until we are able to feed ourselves. If we don't have a parental relationship in our lives, this can make us feel insecure.

As we develop through puberty, this chakra stimulates the endocrine glands of our sexual organs and channels this energy into our reproductive system. This chakra connects to our basic needs such as shelter, food, sex, water, clothing, feeling safe and having a sense of belonging. The energy travels from the base of the feet, then up to the ankle, calf and thigh, where is it processed. The energy force then travels up through the spine and activates the endocrine system. This chakra is associated with the element of earth.

Activity: *Connecting to your root chakra*
- Go for a long walk in nature and collect gifts from nature that represent the earth element: leaves, flowers, earth and soil.
- Create an altar where you can meditate to enhance this connection – find a quiet area in your home where you can put a small table.
- Place a ruby red table cloth on the table, along with all the gifts from nature you have collected that represent the earth element.
- Add gifts that represent the other three elements: water (a chalice of fresh water or sea water), wind (a wind chime), fire (a brown candle).
- Add any objects that remind you of your childhood, including photos.
- Spend time at your altar, and meditate there for at least 15 minutes a day.

» **Problems**
Overeating (causing obesity and other health issues); being materialistic and money orientated; imbalances of the sex hormones leading to infertility.

» **Characteristics**
Procreating, stability, sensuality, security, grounding, basic instincts, connection to environment and Mother Earth.

» **Affirmation**
I am secure, stable and grounded!

The sacral chakra

This chakra brings subtle energy vibrations into our body – part of the electromagnetic field surrounding the planet. These can be compromised in our modern day of electromagnetic pollution, mobile phone rays, microwave rays and X-rays. This chakra is linked to the flow of liquids in the body, and to our 'fight and flight' reflex, which involves the release of the adrenaline hormone from the sympathetic nervous system in the brain. This hormone is released when we are in a stressful situation, and helps us to switch into survival mode. This chakra helps to balance stress levels and governs our creativity and sexual impulse energy. It is associated with the element of water.

» **Connecting to your sacral chakra**
This is the time to connect to your creative side. Being creative is such an accomplishment for human beings, and it is amazing to know that we can activate this in our everyday lives. Creativity is linked to our hands and feet and connects to our heart. Have you heard the saying, 'Do things that you love?' Sometimes in our everyday lives,

such as in our job, we may have lost enthusiasm – but the joy of being creative can help us feel happy, whether it be through dancing, singing, acting, yoga or painting. Spend at least 30 minutes a day being creative in any way you enjoy.

» Problems

Problematic relationships, addictions, illnesses such as depression and anxiety, violence, anger, fear, sex addiction, irritable bowel syndrome, bladder tumours.

» Characteristics

Feeling, reproduction, creativity, happiness, relationships (love and friendship), pleasure, emotions.

» Affirmation

I am creative, energetic and sensual!

The hara chakra

The hara chakra is activated by the *qi* energetic force, and can be achieved by eastern energy techniques such as tai chi. This connects to the vital energy source that comes from within. Breathing techniques are important to help balance this chakra. It is the root and origin of qi energy; once the chakra is activated, it distributes qi energy all over the body and helps us overcome our fears. This chakra connects to the wisdom of the east, and is activated in people who practise energy techniques such as tai chi, reiki, yoga and eastern martial arts including karate and sumo. In oriental medicine, vital life-force energy is called the hara; when it is balanced, you will be in optimum health! Yoga teaches us that all energy channels originate from the navel as the source of life. In reiki healing, the hara, the heart and the crown chakras are the symbolic three diamonds.

This chakra works on our kidneys, digestion and the absorption of food. The strength that it brings into our body will aid metabolism and weight loss in combination with eastern martial art techniques such as those mentioned above.

» Activating your hara chakra

It is important to ensure all your energy fields, such as your chakras and aura, are balanced; and that all your fears, passions and pleasures are acknowledged. To activate this chakra, take classes in eastern energy techniques such as tai chi, and take a yoga class.

» Affirmation

I am energy! Energy is everywhere and everything!

The solar plexus chakra

This chakra is located in the navel and is sometimes referred to as the 'gem of the navel'. It is our connection to the power and energy of the sun, known as solar consciousness. This chakra is activated by solar light energy, which brings these energies in our body, and by the element of fire. This activates the nerve bundles that radiate the great solar plexus. This chakra is located in the digestive system, and is involved in metabolic reactions. This helps to overcome nervous tension caused by stress.

The solar plexus is the centre of personal power, ambition, intellect and desire. This chakra contains a protective energy that guards against negative energy that is contained within any of the other chakras. This is why it is extremely important to protect the solar plexus energy, especially if you are working as an energy practitioner such as a healer or psychic. In energy work, the solar plexus absorbs the energy and thoughts of others consciously or unconsciously, which can withdraw your own positive solar plexus energy out and replace it with the negative energy; this causes discomfort in the stomach. When I started working as a healer and practising angelic reiki, on many occasions I would experience severe pain in my solar plexus from absorbing the recipient's energy and taking it into my own energetic vortex. This caused me to feel extremely ill, tired and sick.

(Protecting your energy during energy work is so important: this will be discussed in great detail in the next chapter.)

Activity: *Connecting to your solar plexus chakra*
1. Go out into the sunshine and absorb its energies. Feel powerful and know that you can achieve anything you want to achieve. Relaxation and breathing techniques are so important to keep this chakra balanced. Go for a swim or a walk, or do any other activity you enjoy, at least three times a week.
2. Have a pamper day:

- Firstly, turn off your phone and any other digital devices so you don't get any disruption from work or the outside world.
- Watch your favourite film or TV programme.
- Eat nutritious food, and drink nice tea or water.
- Do a few stretches or some yoga.
- Run a sea salt warm bath with relaxing essential oils including lavender, peppermint, rose and jasmine.
- Listen to some calming music such as sounds of nature.
- Meditate by sitting still or listening to a guided meditation.
- Get a good night's sleep.

I recommend having a pamper day at least twice a month, if not more – you will be amazed by how much your stress levels reduce.

» **Problems**

Many problems can arise if this chakra is out of balance, including digestive problems, diabetes and cancer. Stress is the biggest issue that swings this chakra out of balance, because stress causes many uncomfortable symptoms: headaches, dizziness, aching joints, susceptibility to illness, high blood pressure, asthma, excessive sugar in the blood, sexual dysfunction and hormonal imbalances. Mental and emotional issues can also arise such as fear; intimidation; loss of self-esteem, self-confidence and self-respect; loss of care for oneself and others; self-destructive behaviour; and sensitivity to criticism.

» **Characteristics**

Digestion, power, expansiveness, growth, metabolism, zest for life, ambition, confidence, intelligence, thoughts, desire.

» **Affirmation**

I am powerful! I am strong! I know who I am! I am amazing!

The heart chakra

This chakra is located in the heart and is the centre of the body – the bridge between the lower chakras and the upper chakras, including the newly emerging chakras. The heart chakra is associated with the circulatory system: hollow muscle that pumps oxygen-rich blood from the lungs to all parts of the body. The blood has three main functions: conveying food, water and oxygen; carrying all forms of energy; and removing waste products that are excreted. The heart chakra governs the immune system; it spins faster during infections such as the common cold, and brings healing to the throat. It is the centre of love, group consciousness and spirituality. It associates with oneness and the element of air.

Activity: *Connecting to your heart chakra*
1. The first way to connect is to have love and compassion for yourself; otherwise you will not be able to emit this loving energy to others.
2. Self-healing:

- Find a nice quiet place, somewhere where you will not be disturbed.
- Have a large, clear quartz crystal in your hand (if you work as a healer, you will probably have one of these; otherwise you can buy one from a crystal store).
- Sit cross-legged on the floor with your crystal in both hands.
- Close your eyes and say a little prayer; say why you need healing at this time.
- Call in your guides and angels to assist with the healing.
- Sit in this space for 20 minutes.
- Write about your experience and how you are feeling; pay close attention to the emotions you are experiencing. Do you feel calm? Do you feel relaxed? Do you feel loved?

» Problems
In the body: heart disease, heart failure, high blood pressure, valve blockages, heart attack, asthma, breast cancer, emotional difficulties such as abuse (physically and mentally) and lack of love.

» Characteristics
Well-being, gentle love, healing, compassion, unconditional love, circulation, passion, devotion.

» Affirmation
I am loved! I am worthy! I am healed!

The higher heart chakra

This chakra is located just above the heart chakra and represents unconditional love. It works alongside the heart chakra to bring even more love, compassion and healing to this planet, and to the stars, other planets and the universe. The higher heart chakra connects you to the power of forgiveness and helps you to speak from the heart (the language of love, compassion and truth). The vibration of love is the highest you will feel; children and animals can bring instant healing, as they only operate on this vibration!

This chakra connects you to higher consciousness and spirit; it opens up your heart chakra to radiate love to the whole planet, animals and other planets. This will help to raise the vibration of the earth and light consciousness so the planet can be free from fear and despair.

» **Activating your higher heart chakra**
- You can do this firstly through forgiveness to anyone who has caused you pain in the past; transmute the pain and upset energy into love and well-being, and send this out into the universe.
- Show compassion and empathy to others.
- Send healing to others (people, children, animals, the earth and the elements) in the same way you did above in the self-healing exercise.

» **Affirmation**
I forgive! I am love!

The throat chakra

This chakra is located in the neck, close to the throat. It is the centre of communication in all forms: verbal, physical, mental and in a creative form (people communicate through their art, voice and movement). This chakra is all about harmony and balance. It is associated with the respiratory system and the thyroid/parathyroid system: these systems bring balance to the other systems in the body. This chakra is associated with intelligence and the ether element (infinite space or atmosphere).

» **Connecting to your throat chakra**
- Breathing exercises.
- Chanting.
- Listening to sounds.
- Participating in a sound healing session.

(Singing my favourite song in the shower is my favourite way to express myself!)

» **Problems**
Thyroid problems such as an overactive or underactive thyroid; sore throat; coughing; loss of voice; mouth ulcers; respiratory problems; weight issues; dental problems.

» **Characteristics**
Personal expression, following one's dream, decision making, breath, sound, harmony, communication, will, intelligence, independence, security.

» **Affirmation**
I speak my truth! I can communicate in every way!

The third-eye chakra

This chakra is located in the middle of the brow (it is sometimes actually called the brow chakra). It is the centre of psychic power, intuition, energies of spirit, personal magnetism, light and clairvoyance. This chakra helps to balance sleeping and waking patterns, and brings spiritual energy to the physical body via the pituitary gland. This chakra is associated with our five senses: sight, hearing, smell, touch and speech. It helps us to enter the realms of knowledge and wisdom, and unlocks psychic abilities including telepathy, clairaudience and clairvoyance. This chakra is associated with joy, consciousness and reality and with the element of ether.

Activity: *Connecting to your third-eye chakra*
1. Activate it by getting enough sleep, eating healthy food, relaxation time, good relationships and a fulfilling work life.
2. Guided meditation to activate your imagination and visualisation skills:

- Find a nice comfortable place to meditate – indoors or outdoors.
- You may wish to use your favourite essential oil in a diffuser, play your favourite relaxation music and hold your favourite crystal (activating the senses is a great way to connect to or activate this chakra).
- Sit comfortably with both feet flat on the ground.
- Close your eyes.
- Imagine that beneath your feet are many roots, like those of a tree; the roots go deeper down until they are attached to a ruby red crystal. This is called *grounding* your energy.
- Imagine your energy climbing back up the roots and reaching your third-eye chakra.
- You see a rainbow bubble; you climb inside and to your surprise there is a fluffy seat made of a white cloud. You take a seat, and you see a button that says 'My favourite place'.

- You press this button and the rainbow bubble takes you flying across the sky until you eventually reach your favourite place; this could be a forest, a beach or a mountain.
- You feel happy and relaxed in your favourite place.
- Spend at least 20 minutes there.
- When you are ready, climb back into the rainbow bubble; you see a button marked 'My quiet spot'.
- You press this button and before you know it you are back in the present in your home or outdoors.
- Write your experience in your journal.

» Problems
Brain tumour, neurological disturbances, seizures, learning difficulties, headaches, sleeplessness, hormonal imbalances.

» Characteristics
Intuition, inner sight, harmony, clarity, imagination, meditation.

» Affirmation
I can see clearly with my inner eye! I am aware! I am intuitive!

The crown chakra

This chakra is located slightly above the head. It is the gateway to spiritual connection and connects to the brain and the nervous system. It combines the energy of the masculine (yang) solar energy and feminine (yin) lunar energy, and symbolises the balance of duality within us and our ability to experience super-consciousness and cosmic consciousness. It focuses on the energetic information coming through the auric field at the crown of the head. This chakra unlocks the mind through meditation in three stages: focusing on the experience, expanding on the experience and deepening the experience. It connects the mind, body and soul in a cycle of complete balance, and is associated with the element of spirit.

» Connecting to your crown chakra

A good way to do this is to exercise your mind by doing crosswords or sudoku, learning a new language or studying something new (the more you exercise your mind, the more you will connect with the crown).

» Problems

Energetic disorders, including chronic exhaustion that is not linked to a physical disorder; extreme sensitivities to light, sound and other environmental factors; headaches; epilepsy; paralysis; neurodegenerative diseases such as Alzheimer's and Parkinson's (a neurodegenerative disease is progressive degeneration – the death of nerve cells in a specific part of the brain that can control movement or thought).

» Characteristics

Meditation, universal consciousness, spirituality, enlightenment, dynamic thought, energy.

» Affirmation

I am present! I am balanced! I am connected to my spirit, body and mind!

The causal chakra

This chakra is located 10cm above the head. It helps with our spiritual exploration in this life plane. It exchanges energy with the guiding higher consciousness, and helps us to receive guidance from the spiritual realms. This guidance is received when we open our crown chakra and connect to spirit. It is possible to do this by guided meditation, and by using scrying tools, crystals and tarot/oracle cards. This chakra opens us up to being more aware of our soul group; our soulmates can be incarnated on the human plane at the present time, or be in spiritual dimensions. This chakra is associated with karma in the physical lives. Karma is the cause and effect that happens as part of the action we take; it is like spiritual baggage we have chosen to carry, but also could be what we want to release in this life.

Activity: *Activating your causal chakra*
To activate this chakra, you must release everything that is not serving you any more by performing a ritual:

- On a full moon, write on a piece of paper what you want to release: for example, I release all negative people who are attached to my aura and drain my energy (unconsciously and consciously).
- Once your list is completed, you must burn it under the moonlight, preferably outdoors.
- Write a positive version of your list: for example, please surround me with people who are positive and value me for who I am.
- Keep this list; on the new moon, manifest the energy that you truly want.

(On a full moon, it is a good idea to cleanse and recharge your crystals directly under the moonlight.)

» Affirmation

I am connected to my soul and soul group! I am open to receiving guidance from other dimensions and realms!

The soul star chakra

This chakra is located around 15cm above the crown chakra. It connects us to the Milky Way, which transmits energies to and from an awakened soul. Healers who heal humans and animals are aligned to this chakra fully where they are capable of channelling this divine energy from above and beyond. A healer is attuned to certain frequencies and symbols, and has the intention to heal to bring maximum benefit to the recipient. A healer can be referred to as a 'lightworker' because of their understanding that light can raise the vibration of human consciousness.

This chakra is associated with personal and planetary healing. It is the chakra needed to activate DNA codes (remember DNA is where all our genetic information is stored – what we have inherited from our parents). Once awakened, this chakra stimulates dormant codes that open up new talents and gifts we have within our make-up.

DNA is activated through unidentified quantum energies that work in a multidimensional way. DNA activation can be initiated by a qualified healer who specialises in a healing practice that involves quantum physics. Quantum physics uses the energies from the earth and other planets to activate dormant parts of our DNA that have been waiting to be activated, to enhance our experience and unlock our skills and talents in this incarnation on this planet.

The other role of the soul star chakra is to help the soul to depart after death of the body. The soul can take time to fully leave the earth, especially if the person has died suddenly or in an accident, or even been murdered; sometimes the soul hasn't fully comprehended that it is no longer in a physical body. A trained healer can help in the transition of the soul back to the spiritual realm, and assist it in its ascension.

» Activating your soul star chakra

This can be done by mindfulness meditation, and visualisation of all the seven main chakras being in balance and in harmony. Mindfulness meditation will be discussed in greater detail in Chapter 8.

» Affirmation

I am the driver of my life journey! I am balanced and in harmony with myself, the earth and all the planets in the galaxy!

Stellar gateway chakra

This chakra is located roughly 30cm above the crown chakra, and is the third newly emerging celestial chakra. The causal, soul star and stellar gateway chakras are our connection to our soul and spiritual self. The stellar gateway chakra illuminates awakened humans with the divine light of all creation. To prepare for this chakra, you must act compassionately, bring unconditional love and be non-judgemental. It is linked to the newly emerging higher heart chakra.

This chakra is your access to other planets, galaxies, universal energies and spiritual realms beyond our physical understanding. It is our primary connection to other realms – like a doorway that takes you well beyond the reality you are in now. It is the gateway to other living beings in the sky, planets, the stars and beyond; angelic realms, elemental realms (dragons, mermaids, fairies and unicorns), star beings, extraterrestrial beings.

Activity: *Activating your stellar gateway chakra*
Stargazing is a great way to work with this chakra:
- Go out on a clear night where the stars are shining so brightly and the moon is visible; this could be in your garden, in a park, in woods or on a beach.

- Find a nice spot and sit down cross-legged; open your hands with the palms facing the sky.
- Stare up into the sky and observe all the beautiful stars and the moon – if you are lucky enough, you may see another planet, or even a shooting star!
- Say a little prayer asking to connect to the universe.
- Concentrate on your breathing and close your eyes.
- Imagine yourself on a journey in a rainbow-coloured bubble. You are floating around the universe. Take note in your mind of any colours, shapes or even beings that are visible to you at this time.
- Stay in this space for about 20 minutes, and when you are ready, return to your physical self.
- Open your eyes, take a few moments to come back properly, then look up at the stars and the moon: do they look different now? How do you feel? Did you meet any spiritual beings or ETs?
- Write this all down in your journal. The more you do this, the more your stellar gateway will be activated and awake.

» Affirmation

I am a star being! I am love! I am light! I am a vortex of energy!

The endocrine system and the chakras

It is so important that these energetic systems work together on a physical and spiritual level. The endocrine glands in the endocrine system are in charge of producing and regulating hormones, and making sure they are sent to the right area in the body to regulate a specific function. They work alongside the chakras to regulate and, most importantly, bring balance to the body, mind and soul in every way.

Atlantis

» My dream

The reason I was called to write about Atlantis is that last night, 11th November, I had a dream. In my dream, my dad was there as a spirit guide with his council of elders. We were discussing karma and protection; my dad advised me to write on a piece of paper all the people and situations that caused me harm, and say a prayer to rid me of these energies and transmute the negative energy into positive energy to send back to the universe. I said I would do this on the next full moon, which, funnily enough, was happening on the following night! My dad said to me: 'Don't forget to take your dolphin with you,' and I said, 'OK, Dad, I will – see you later.' It was so strange – it was my spirit guide dolphin that has been with me since Atlantis; he is my teacher, guide, healer and protector, just like my father was before he died 6 months ago.

When I woke up in the morning, I had to hold my green marble dolphin; I sprayed my mermaid aura spray, and went into a meditation with the sounds of the dolphins. This took me to a familiar place of peace, wisdom, love and joy (all characteristics of my spirit dolphin). The other characteristics of a dolphin are power, strength, knowledge and speed to complete any task; in my case it is completing the manuscript for this book.

» What was Atlantis?

Atlantis was an ancient civilisation that existed thousands of years ago. Plato (a Greek philosopher living in Athens) was the first person to write about it, in 355 BCE. Atlantis was around 9,000 years before this. It is thought to have begun in Greece and extended to other places within the Atlantic (hence how this ocean got its name), such as the Canary Islands, close to Africa.

The Atlanteans were described as highly intelligent, technologically advanced (using crystals), wealthy, attractive and, most importantly, spiritually advanced (using nature gifts such as crystals, flower essences, and essential oils for health and healing).

Atlantis was a sea kingdom containing a vast number of islands. The main island and capital was the city of Poseidia; within the city were large temples containing crystal computers that radiated information across the Atlantean empire. Crystals have strong electromagnetic fields (clear quartz crystals having the strongest) with different vibrations; this is not so different to the modern world, where crystals are needed for most technological equipment to work. Small, clear quartz crystals are found in every electrical circuit; these are needed to transmit electricity. You can buy watches powered by clear quartz crystals. Phones, televisions and music systems are all powered by crystals (crystals will be discussed in detail in Chapter 6).

Atlantis had many pyramids, which were used to generate energy for great spiritual technologies. This could be why ancient civilisations after Atlantis, such as the Egyptians, built temples for spiritual worship and prayer, and to connect to their gods and goddesses. So why did the Atlanteans build pyramids? Their idea was that at the base of the pyramid there were libraries of energy centres, where people used to study or even live! This was their connection to earth; I expect this is where their earth star chakra was activated and they could communicate with other galactic societies such as the Pleiadians, Sirians and Archurians. The tip of the pyramid was to connect to the divine creator energy, light and higher consciousness – like the crown, causal, soul star and stellar gateway chakras in the chakra system.

» My connection to Atlantis

My father was of Greek descent; he was from a small Greek island called Kimolos, which is part of the group of islands called the Cyclades enclosed by the Aegean Sea; it is a very small island with roughly 600 inhabitants. The Cyclades are islands that form in a circle and include other islands such as Milos, Santorini, Mykonos and Delos, which is the central island of the Cyclades. Delos is one of the most important mythological sites in Greece, containing ancient stones going back to 300 BCE (interestingly, the same time that Plato wrote about Atlantis). It is an island of history, Greek mythology, war, archaeology and spirituality. It is the birthplace of the gods Apollo and Artemis, and is seen as the sanctuary of Zeus (king of the Greek gods).

If you take a boat from Kimolos to the neighbouring island of Milos, you may see in the ocean the remains of ancient temples: pillars, pottery and artefacts (remains of ancient Athens or possibly even from Atlantis). Most of my father's family moved to Milos to set up a life there; others moved to Athens. As a child, I would spend my school summer holidays in Milos. I loved it, and, more importantly, loved swimming in the sea. I always felt such a strong connection there, like I had been there before. Many archaeological relics have been found in Milos, such as the Venus de Milo (ancient statue of the goddess Aphrodite) and statue of Poseidon (the Greek god of the sea, storms, earthquakes and horses); his son Atlas was said to be the first king of Atlantis, and Atlantis was Poseidon's domain. Also, the statue of Apollo was found in Milos – he is the Greek god of many things: the sun, light, oracle, knowledge and music, along with being the protector of the young and much more. I am sure he had an influence on Atlantis and taught the Atlanteans so much.

So, Atlantis was this super-duper place of knowledge, wisdom, advanced technology, wealth, abundance and fruition! And so we may ask: what went wrong? It was **greed** – the negative ego coming in, too much wealth, advanced technology and power got to the Atlanteans' heads. They started misusing their power for personal gain, and treating others cruelly. They enslaved the hybrid beings they had created, punishing them and making them go through a purification process – these hybrids were of animals, humans and other beings. They treated the Lemurians very badly. The Lemurians were beings from Lemuria that took refuge in Atlantis when their home was destroyed; they brought the Atlanteans gifts of love, light, compassion and crystals. The Atlanteans thought the Lemurians were not pure, and would make them go through a horrific purification process, before enslaving them. The Atlanteans did not just enslave others: they would switch their DNA off so they could control them in every way possible. This is interesting, as we are now aware of DNA activation techniques (as discussed above) that help to switch the dormant DNA on and birth our knowledge and skills, open up our consciousness, ideas and creativity, and overcome the negative ego that drives fear, despair, greed and depression.

The Atlanteans didn't just hurt people physically, but also sent out negative thoughts to control them and shut them down. It must have been so awful! This still happens in the present day where people use abuse to make them feel powerful, to control others and to make them feel inadequate in society. Environmental energy, such as in the workplace, is similar to this, where people exert their negative energy and thoughts on others. I felt this strongly in my previous workplace, and it made me feel uneasy every time I was in that environment. I would get ill very easily with colds and headaches, and felt really tired. It got so bad that eventually I had to leave. This is why it is so important to protect, ground and balance your energetic aura; techniques for doing this will be discussed in the next chapter.

The Atlanteans lost their spirituality and connection to the divine energy of creation, and eventually Atlantis collapsed and was covered by the ocean.

» My memories of Atlantis

I am a trained angelic reiki practitioner, and during my training I was given an angelic reiki healing from a fellow student. During the healing, I got a few great visualisations: I saw myself on a boat in the Mediterranean Sea, floating along; then this beautiful dolphin climbed into the boat and stood upright like a human and spoke to me: 'Be careful on your journey.' He kissed my cheek, then jumped back into the sea. Shortly after that, Poseidon rose from the sea and said, 'Please, my child, be careful of the sea. The energetic aura of the sea can change suddenly – it is not always calm. I will always protect you!' Then he disappeared.

My first contact with the ocean was when I was a baby: my father threw me into the sea and I started swimming. Since that time, I have always loved being in the ocean; my father would always call me his little mermaid – 'γοργονα' in Greek. As a child, I would spend hours in the ocean connecting to the energetic aura of the ocean, and of the marine and ocean plants. My father also was in the ocean a lot and even worked as a sailor. I feel that he would have been an Atlantean like me. I feel such a strong connection to the energy of Atlantis. I love going swimming regularly, especially in the sea.

» The magical beings of Atlantis

Atlantis was the home to many magical beings of light, such as the fairies, unicorns, merfolk, dragons, hybrids, shapeshifting beings and, of course, the Greek/Roman gods and goddesses. This may be the reason why I have such a huge fascination with these beings: I enjoy reading and studying them, I use oracle cards to channel their energy and I listen to guided meditations to meet them; I must be remembering my time in Atlantis and being with these magical beings. I enjoy visiting sacred sites where they have existed, such as Glastonbury (fairy realm), the Acropolis in Athens (Greek gods and goddesses) and the Atlantic Ocean (merfolk and Atlanteans). All these magical beings, including animals, had great talents and gifts that they offered to the Atlanteans. In Atlantis, animals were treated as equals with love and respect; they were in charge of large learning facilities, and Atlanteans even went to school with their cat or dog (this may explain why humans in the present time have so much love for their pets and treat them as part of the family). The centaurs (half human and half horse) were extremely intelligent, as they had a human brain and taught subjects such as medicine, mathematics and science to the others.

Activity: *Creating an altar to connect to the magical beings of Atlantis*
- Gather pictures and ornaments of the mythical beings you want to connect with: for example, pictures of mermaids, dolphins or fairies.
- Place objects that you associate with these beings, such as seashells, flowers and crystals (use larimar to connect with the mermaids) on the altar.
- Listen to the sounds of nature, or download a guided meditation connecting to Atlantis.
- Record your experience in your journal.

» The dolphins of Atlantis

Dolphins are magnificent creatures and highly intelligent; they contain a part of the brain similar to a human brain, called the *neocortex*. This is the part that is involved in higher brain functions such as sensory perception (see Table 3.5), language development, cognitive function and body movements. The neocortex in a dolphin's brain is large, and has a great number of neocortex neurons (neurons will be described in detail in Chapter 8 on the brain and mindfulness). This could explain how dolphins interact with each other similarly to the way humans do: they live in affectionate family circles called *pods*; they are very sociable and like to interact, play and communicate through sound, even with humans. Each dolphin has their own unique high-pitched whistle, similarly to humans who each have their own unique voice.

Humans perceive their environment through sight; so do dolphins, but they also use a technique called *echolocation* to locate objects in their environment. This works by the dolphin sending out calls to the environment and listening to the echoes of the calls that return from the objects around them. Pretty cool, right? We can communicate telepathically with dolphins, and they have the ability to understand us.

Dolphin energy unites Lemurian consciousness with Atlantean technology. Dolphins open you up to increased frequencies of light and other dimensions, because they contain Sirian energy and can help humanity evolve into beings of light. Dolphins help the crystalline gridding system that surrounds the earth; they resonate with sacred geometry (use of symbols) and frequencies of light to transmute our physical body into a body of a higher vibration – the vibration of light energy.

In many ancient civilisations, including Atlantis and Greece, dolphins were considered sacred animals and our equal. For example, in ancient Greece, if you killed a dolphin you would face the ultimate punishment and be killed!

Dolphins were regarded as messengers from a god/goddess or the creator. Aphrodite, daughter of Poseidon, had two soul dolphins that were always with her, a little like my dolphin spirit guide. In Atlantis, dolphins were leading guides and teachers/educators in oceanic school, and would communicate by sound as they do now. Dolphins were respected because of their beauty, courage, good humour and strength.

Activity: *Connecting to the dolphins of Atlantis*
- Find a nice place to sit, preferably by water – the ocean, a river, a waterfall or even a small stream; or you could also do this indoors by your Atlantis altar.
- If you have mermaid and dolphin pictures and ornaments, place them next to you.
- If you are not near water, play the sounds of the ocean; this will enhance your connection.

- Now I want you to imagine that you are swimming in the aquamarine ocean – you are feeling so happy and relaxed. The sun is shining brightly, and the water is so still and calm.
- Your attention is drawn to the ocean in front of you; it goes straight ahead towards the dark blue / deeper ocean, and you see a pod of dolphins. They are starting to swim towards you; as they get closer to you, the water starts making waves.
- The dolphins come closer and kiss you on the nose! This brings so much love, joy and happiness to you.
- There is one particular dolphin you connect to the most, and he/she wants to communicate with you. Does your dolphin have a name? Does your dolphin have a message for you?
- Spend some time with your dolphin, and when you are ready, open your eyes.
- Write about your experience in your journal.

Reading

Anshara S. *The Age of Inheritance: The Activation of the 13 Chakras*. QuantumPathic Press, 2003.

Cavendish L. *The Lost Lands: A Magical History of Lemuria, Atlantis and Avalon*. Blue Angel Publishing, 2013.

Mercier P. *The Chakra Bible: The Definitive Guide to Working with the Chakras*. Octopus Publishing Group, 2009.

Mercier P. *The Little Book of Chakra: Balance Your Subtle Energy for Health, Vitality and Harmony*. Octopus Publishing Group, 2017.

Pond D. *Chakras Beyond Beginners: Awakening to the Power Within*. Llewellyn Publications, 2016.

CHAPTER 5

Grounding and balancing the chakras

Do you often feel drained when returning home from a day out? Do you feel overwhelmed and lack energy in the presence of certain people? Do certain noises and smells affect your energy or mood? Are you sensitive to artificial light and laser beams? Do crowded, noisy places leave you feeling exhausted?

If you have these characteristics, you may be an *empath*. What is an empath? An empath has the ability to feel and sense negative energies of others who are in close proximity to their energetic aura. An empath is sensitive to noise, light, smells and a negative environment. This is why it is so important to ground and balance your energy!

As discussed earlier, energy is everywhere. Every living thing has an energy attached to it.

Energy cannot be made or destroyed, it can change from one form to another.

Albert Einstein

The key thing in this quote is that negative energy can be transformed (changed) into a positive energy.

There are three important ways of achieving this:
- **Protecting your energy** – this will help you prevent negative energies entering your field and attaching to your chakras, making you feel out of balance and unwell.
- **Balancing your energy** – this is so important, as it will help change the vibration of your auric field, making sure you don't experience any blockages, your energy flows freely throughout the body and all your chakras are vibrating at the right speed.
- **Grounding your energy** – connecting to your root chakra will help you feel more in control of your energy, and you can use this energy to help ground the other chakras.

I am writing this chapter from my own experience and how I have learned to keep my energy protected, balanced and grounded.

Protecting your energy

I must protect my energetic aura and chakra system from any external energies from other people, smells, noise and so on. My aura needs to be protected at all times because of the work I do. I need to protect it from the public, as they seem to be attracted to my aura and want to attach their energy to it; most of the time it is unintentional, but it nevertheless makes me feel uneasy.

» Best ways to protect your energy
- Imagine a large blue cloak of protection being placed around your body; any unwanted energy that tries to attach itself will be repelled and will bounce off the cloak like a shield.
- Surround yourself with bright white light!
- Say a prayer to your guides and angels to protect your energy from energy vampires.
- The colours of protection are blue, violet and white. These colours

operate on a high vibrational frequency, which is needed when you need protection. Wear these colours, and buy crystals of these colours to carry with you when you are out.

- Carry black crystals, such as black tourmaline and black obsidian, with you when you are in crowded, noisy places. Both of these crystals will help to repel negative energy trying to come into your energetic field. I have a black tourmaline crystal keyring for my keys, to help protect my energy.
- Before you leave the house, say a little prayer to your guides and angels to protect your auric field and keep you safe.
- Release any fears you have through meditation and prayer to stop negative energies manifesting.
- Say this affirmation three times: 'I am safe and totally protected by the universe from all energy that is not for my highest good.'
- I always connect to Archangel Michael when I need to be protected. I close my eyes and imagine Michael there in his armour, giving me a sword and shield to protect me against any unwanted energies. I imagine his large white wings wrapped around me, protecting me from anyone wishing to do me harm. I see Archangel Michael as my spiritual guide and father. He can protect you too – all you have to do is ask!
- Release any energetic ties you have to others who drain your energy and make you feel unwell. Archangel Michael can also help with cord cutting and detaching the energies that are not doing you any good.
- Protect your space: burn sage sticks, burn frankincense and use singing bowls or chimes to get rid of any unwanted energy in your home. Whatever you do, makes sure you protect all corners of the room.

Balancing your energy

- Go out in nature and connect with all four elements: earth, wind, fire and water. The best place will be a park where there is a stream or lake, sunshine, wind and lots of grass with trees.
- A great way to rebalance all the chakras is to go on holiday and have pure relaxation, away from the hustle and bustle of city life, work life and constant use of technology.
- Spend a week or two in a quiet and tranquil place where you can connect to the four elements (earth, wind, fire and water), the animals, the moon, the sun and nature.
- Find time to meditate and connect fully with yourself – physically, spiritually, mentally and emotionally.
- It is a good idea to journal your experiences so you can identify activities that help you to feel energised, happy and rested in your day-to-day environment.

I really enjoy going to the Canary Islands because whatever time you go in the year, it is lovely and warm! I find these islands really peaceful, and I love to connect with the mountains and wildlife and to swim in the ocean. This truly helps me feel balanced. I enjoy meditating as the sun rises and as it sets. Both such beautiful energies! I enjoy watching the moon rise too, as this brings peace to the end of the day.

» Chakra balancing

A key aim of this book is to ensure you can identify when your chakras are over-stimulated, balanced or under-stimulated. This is so important, so you can bring these chakras back into balance to feel happy and well. Chakras out of balance can lead to certain negative behaviours, and even illnesses such as depression and anxiety. When they are in balance, you will feel well and energised in that part of the body.

Over-stimulated chakra: This is when the chakra is working overtime, which can bring about unwanted characteristics such as anger.

Under-stimulated chakra: This is when the chakra is not working to its full capacity and is lacking in activity. This can also have negative effects on the body such as constant tiredness and loss of confidence.

To understand whether your chakras are out of balance, please refer to Table 5.1. It focuses on the seven main chakras of the body – the root, the sacral, the solar plexus, the heart, the throat, the third eye and the crown – and describes characteristics of an over-stimulated, under-stimulated and balanced chakra.

Table 5.1: Balance of the chakras

Chakra	Over-stimulated	Under-stimulated	Balanced
Root	Materialistic, aggressive, egocentric, picks up negativity, unrealistic	Insecure, lack of confidence, low willpower, no interest in self-care, depressed, not grounded	Grounded, at one with nature, the earth, the elements and the animals

Secure, nurtured, all basic needs fulfilled, practical |
| Sacral | Overemotional, sex addict, overindulgent, manipulative, selfish, angry | Oversensitive to every stimulus

Resentment, guilt, distrust, staying in abusive relationships, sad, infertile | Stable relationships with a balance of giving and receiving

Creative, sexual energy balanced |
| Solar plexus | Judgemental, workaholic, perfectionist | Loss of confidence, loss of self-esteem, stressed, fearful, poor digestion, exhaustion | A good balance of ego, confidence, digestion, rest, metabolism and relaxation

Good energy levels |

Heart	Demanding, moody, disconnected from feelings, jealous, possessive, controlling, unable to show love	Indecisiveness, fear of rejection, in unloving relationships	Generous, compassionate, giving unconditional love, helping others, empathetic
Throat	Arrogant, talking excessively, not listening to others	Not good at communicating, incapable of verbalising thoughts or feelings, feeling stuck in the wrong situation	Expressing yourself in every way Communicating freely by speaking and writing Listening and paying attention to others
Third eye	Overly proud, manipulative, untrustworthy, attached to past	Unable to tell the difference between the ego and higher self, head in clouds (spaced out) all the time, oversensitive, non-assertive	Intuitive, thoughtful, mindful, perceptive, visionary
Crown	Frustration, frequent migraines, headaches, overly imaginative	Feeling lost in society, depressed, sad	Imaginative, aware of spirituality, creative, humanitarian

A great way to balance your chakras is by a process called *colour breathing*. You can do this indoors or outdoors.

Activity: *Colour breathing*
- Close your eyes and take a few deep breaths.
- Start at your root chakra; imagine a large red rose opening up at your root; breathe in the colour red and say, 'I am grounded in a bed of red roses.'
- Move up to the sacral chakra; imagine a beautiful large orange smelling so sweet there at the sacral;

- breathe in the colour orange and say, 'I am in a beautiful field of oranges.'
- Move up to the solar plexus; imagine a large yellow sunflower opening up there; breathe in the colour yellow and say, 'I am a bright yellow light in a field of flowers.'
- Move up to the heart chakra; imagine a large emerald crystal shining in the shape of a heart; breathe in the colour green and say, 'I am the green field full of love.'
- Move up to the throat chakra; imagine a bluebell flower opening up; breathe in the colour blue and say, 'I am as blue as the midday sky.'
- Move up to the third-eye chakra; imagine a bush of lavender flowers opening up; breathe in the colour violet and say, 'I am a violet flower in a field of many.'
- Move up to the crown chakra; make a triangle shape above your head and imagine bright white light like a feather from a dove opening up; breathe in white light and say, 'I am a pure white light emitting energy to the sky.'
- Say, 'I am all the colours of the rainbow.'
- Take in a few more deep breaths.
- Repeat the sequence above twice.
- Then **relax**!

Grounding your energy

The main way to ground all your energies is to connect to the earth and the environment. You must connect with all four elements – earth, wind, fire and water – and connect with your mind and soul. You also must connect to the seasons: winter, spring, summer and autumn.

» The elements

Connecting to the earth element: Go outside in a park or wood where there is an abundance of trees. Close your eyes; in your mind, say three times, 'I am rooted into the depths of the earth like a tree – I am the link between the earth and the sky'. Concentrate on your breathing; each breath you take is earth energy and your connection to the earth. Spend time observing nature: hug a tree if you wish (if you listen closely, you may hear the fluid of the tree rushing up and down), smell the flowers, listen to the sounds of the birds and take time to be silent in your mind.

Connecting to the wind element: Go up on a hilltop and listen to the sounds of the wind. Look above and see the birds flying; without the oxygen that is in the wind, this would not be possible. Concentrate on your breathing – without that oxygen, we and the other living beings on this earth would cease to exist! Breathe in and out in rhythm with your body and say in your mind three times, 'Every breath I take, it is because of you – I am grateful for this source of life.'

Connecting to the fire element: Go outside when the sun is shining brightly in the sky. Take a seat on the ground and close your eyes, feel the warmth of the sun's rays on your skin, and say in your mind three times, 'Thank you for the warmth you give me and all the vitamin D needed for my body to function.'

Connecting to the water element: Go for a swim outdoors, pay attention to the rhythm of the water and say in your mind three times, 'I feel purified, vitalised, relaxed and calm, and it is thanks to your energy.' Drink at least two litres of water a day to feel hydrated and healthy.

» The seasons

Connecting to the season of winter: Wrap up warm and go for a long walk in the snow, build a snowman, ride a sledge and take time to rest like a bear in hibernation. Say in your mind three times, 'I embrace your coldness and darkness and know it is time to wrap up warm and hibernate.' Rest as much as you can.

Connecting to the season of spring: Go outdoors and see all the baby animals being born. Say in your mind three times, 'I am grateful for all new forms of life.' Spend time in nature and see how wonderful the cycle of life is.

Connecting to the season of summer: Spend hours outdoors; watch the sunset and sunrise. Say in your mind three times, 'I am blessed with vibrant light, energy and warmth.' Go for a swim in the ocean, sunbathe (make sure you wear skin protection) and go and pick your own fruit and vegetables.

Connecting to the season of autumn: Go outdoors and observe the falling leaves of different colours: yellow, orange, brown, red and green. Take some time to run and jump in them. Say in your mind three times, 'I am ready for transformation and change – thank you for showing me this.'

Activity: *Recording the balance of your chakras*
Record whether your chakras are balanced, over-stimulated or under-stimulated in Table 5.2.

Table 5.2: Recording the balance of your chakras

Chakra	Balanced	Over-stimulated	Under-stimulated	Comments
Root				
Sacral				
Solar plexus				
Heart				
Throat				
Third eye				
Crown				

Spend the next few weeks trying to balance your chakras through techniques described in this chapter and in Chapter 4.

I hope, by now, you are confident and understand how your auric field and chakra system works. It is now time to learn to use nature's gifts, such as crystals and essential oils, to enhance your well-being in all ways possible. You will learn how crystals work and how they can be used to heal the body, and you will learn the magic of essential oils and how they can help you feel balanced. Finally, you will learn why meditations and mindfulness are important in managing your mental health and reducing your stress levels!

CHAPTER 6

Crystals and healing

What is a crystal?

The word 'crystal' comes from the Greek words κρύσταλλος (*krustallos*), meaning ice/rock, and *kruos*, meaning cold. Once a crystal has been formed, it is in the shape of a fixed solid (all atoms are extremely close together with minimal vibration) in a crystalline lattice structure. There are tiny grains of crystals formed within the structure – these arise from intense heat and water environments. Scientific evidence has proven that crystals do vibrate and have both piezoelectric (electricity resulting from pressure and heat) and pyroelectric (the ability to generate a temporary voltage when heated or cooled down) effects. This is why clear quartz crystals are found in most electrical devices: watches, computers, washing machines, lasers and so on. Crystals are able to work through the human energy system: the chakras and aura.

Crystals were used in many ancient civilisations, and there are texts in Chinese, Latin and Greek explaining their magnetic effects from 5,000 years ago. The Greek philosopher Theophrastus wrote a text explaining the taxonomy of known gems, including their origin, and their physical and healing properties. This text is the basis of modern-day scientific classification of gemstones (crystals). Theophrastus was often referred to as the father of botany for his outstanding work with plant classification.

Crystals are found all over the world, and specific crystals are found in unique locations. For example, the crystal larimar is only found in the Dominican Republic; this makes it extremely precious and expensive. Most crystals are found in rocks and can be retrieved quite easily. Crystals are natural solids made from minerals; they are formed in the earth's surface. Each crystal has a precise atomic arrangement, even though they are in different shapes, colours and sizes. Crystals can be identified by their colour, mineral composition, crystal habit and degree of hardness. The degree of hardness of a crystal is described by the Mohs scale of mineral hardness. This indicates how hard minerals are in relation to each other; diamond is regarded as the hardest crystal. Crystals can be bought as polished stones, crystal clusters or raw crystals.

How are crystals formed?

Crystallisation is the process by which crystals are made. The two main ways are:
- Evaporation – where you evaporate the solvent that dissolves the solute (a soluble solid from a saturated solution) by heat; the remaining solute forms the crystals. An example of this is copper sulphate.
- Using a small crystal known as a *seed crystal* in a saturated solution that contains the same substances as the crystal. Over time, the seed crystal will turn into a large crystal. The solution left after the formation of the crystal is called the *mother liquor*.

The particles that make up the crystal can be arranged and bonded in different ways, which is why crystals come in a variety of shapes such as cubic, rhombic, hexagonal, triclinic and tetragonal. Some crystalline structures can exist in two or more different crystal shapes because of a process called *polymorphism*. These changes occur because of changes in temperature, pressure or both. Figure 6.1 shows the different shapes of crystals.

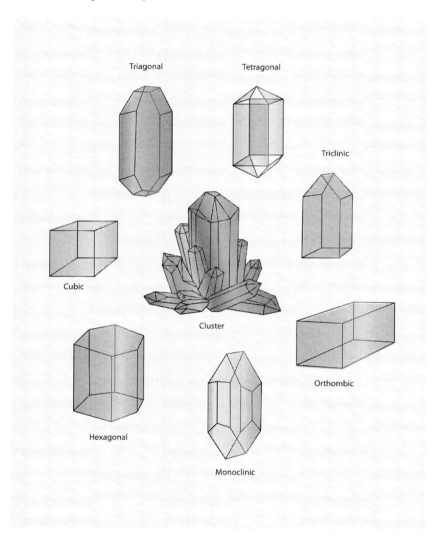

Figure 6.1: Shapes of crystals

Crystal colours

Table 6.1: Crystal colours

Colour	Examples	Characteristics
White/clear	Quartz, white calcite, Dalmatian stone	Healing, purifying, connects to higher consciousness
Black	Tourmaline, obsidian, opal	Protection, repels negative energy and thoughts
Red	Red Jasper, garnet, ruby	Grounding, connects to earth energy, stabilising
Orange	Orange calcite, carnelian, sunstone	Energising, creativity, releasing
Yellow	Citrine, amber, tiger's eye	Power, awakening, stimulating
Green	Jade, aventurine, malachite	Love, calming, balancing
Blue	Blue lace agate, aquamarine, larimar	Communication, calming, intuition
Purple	Amethyst, purple fluorite, sugilite	Spiritual, awakening, humanitarian
Pink	Rose quartz, pink opal, rhodolite	Unconditional love, healing, nurturing
Multicoloured	Jasper, onyx, tourmaline	Balance, grounding, protection
Rainbow	Labradorite, rainbow obsidian, opal	Well-being, balances the spiritual and scientific brain, intelligence
Grey	Hematite, silver, flint	Grounding, connection to metals and earth
Brown	Smoky quartz, rutile, mahogany obsidian	Earth, grounding, detoxifying

Table 6.1 shows the different colours of crystals, with examples and their characteristics. This table will be useful when you are choosing the crystals you need at any given time. It is best to buy a crystal

you connect to rather than choose one because of its type or colour; the characteristics are generic ones and may not resonate with you.

Activity: Experimenting with the energy of a crystal
This simple exercise can give you an experience of working with a crystal. You will be able to see how crystals channel light energy from their environment.
- Get a clear quartz crystal that has a point.
- Hold the crystal in the palm of your hand.
- Point the crystal towards your other hand.
- Slowly rotate the crystal in a small clockwise circle.

As you move the crystal, the point of light moves around your hand.

You may feel a warm sensation in your hand; this is the energy that is coming out of the crystal.

Caring for your crystals

When you have your set of crystals you would like to work with, it is really important to look after them, following the methods below.

» Cleanse your crystals

It is very important to *cleanse* your crystals; this is completely different from *cleaning* them. Cleaning is a physical process, whereas cleansing is an energetic/spiritual one. There are several ways in which you can cleanse and clear the unwanted energy from your crystals:
- Use a smudge stick containing sage (the smoke releases ions that can help reduce stress, so if your crystals have this energy attached to them, sage will help neutralise this).

- Burn incense over your crystal. I suggest using sandalwood – this releases natural antiseptics that will cleanse the crystals.
- Use sound, with a singing bowl, chimes or even a small bell over the crystals; this works by changing the vibration of the crystals to remove unwanted energy.
- Place the crystals in an amethyst bed for a day or more.
- Bury your crystals in the earth on a full moon, and retrieve them on a new moon.

Decide the best cleansing method for you, and do this regularly. Remember to cleanse your crystals after a crystal healing treatment you have performed. I regularly cleanse mine when there is a full moon, as follows:
- Wash all the crystals in warm water.
- Dry them carefully, and place them on a window ledge facing in the direction of the moonlight.
- Spray all the crystals with a special crystal cleansing spray containing sage, frankincense and rose essential oils.
- Leave them on the window sill to absorb the moonlight for three days before putting them away.

» Activate your crystals
- Firstly, you need to attune each crystal to your own frequency. To do this, program your crystal by closing your eyes, putting the crystal close to your heart and saying something like, 'I want this crystal to bring love and healing to everyone it has contact with.'
- Bless your crystal with the highest energy of light from the universe!

(If you have bought a crystal for someone else, it is a good idea to cleanse it and then activate it with the intention you want this crystal to bring to that person.)

» Storing your crystals

- It is important to store your crystals somewhere safe when not in use (remember they have their own vibrational energy, and therefore need to be looked after: keep them clean and clear of dust).
- Keep the crystals you use regularly in a nice silk bag when not in use.
- Have a nice compartmental box to store the crystals you use occasionally.

Crystals and the chakras

Table 6.2: Crystals to place on chakras

Chakra	Crystal	Properties
Crown	Amethyst	Opens up our connection to spirit and our guides
Third eye	Lapis lazuli	Encourages creative expression, wisdom and endurance
Throat	Blue lace agate	Improves communication on all levels
Heart	Green aventurine	Protects, soothes and calms emotions
Solar plexus	Citrine	This crystal is good for relationships and getting rid of emotional toxins
Sacral	Orange carnelian	Improves your mood and enhances creativity
Root	Red jasper	Helps with grounding your energy into the earth

As you have discovered the chakra system in detail in Chapter 4, and have identified the colours each chakra represents, the crystal you will use for a chakra will be the same colour as that chakra. Each colour has a specific vibration; the vibration of the crystal will connect to the vibration of the specific chakra. In my crystal chakra set, I use a particular crystal to place on each chakra during a crystal healing session, as shown in Table 6.2.

Each chakra is, as you know, connected to a specific area in the body and a specific endocrine gland. If this chakra becomes blocked, the energy flow becomes out of balance, which can eventually lead to dis-ease in all areas of the body – physically, emotionally, mentally and spiritual (dis-ease is when your natural state of ease is disrupted).

Crystals to activate the newly emerging chakras

There are numerous crystals you use during meditations to activate various chakras: earth star, hara, higher heart, causal, soul star and stellar gateway.

Earth star: To strengthen your connection to the energy of the earth, use crystals that will ground your energy and have earth-like properties, such as bloodstone, red jasper and hematite.

Hara: To strengthen your connection to eastern energy techniques, use crystals that will enhance breathwork and movement, such as blue lace agate, aquamarine and chalcopyrite – a rainbow-coloured crystal that is good to use when performing martial arts techniques because it improves the flow of chi energy. This specific crystal helps with energy blocks and connects with all the psychic abilities you have, and is a wonderful crystal to use when you are meditating.

Higher heart: To strengthen your connection to unconditional love and compassion, and regulate your cardiovascular system, use any pink crystal (such as rose quartz) or any green crystal (such as malachite).

Causal: To strengthen your connection to higher consciousness and the spiritual realm, use clear quartz crystal or crystals such as angelite, celestine or kyanite. These crystals work well in a meditation when connecting to the spiritual realm.

Soul star: To strengthen your connection to the higher frequency of light and sacred geometry, use white or clear crystals such as selenite, diamond or clear quartz merkabah; or other sacred geometry crystals such as triangular, sphere, cube, octahedron, dodecahedron or icosahedron. A white merkabah represents the soul star chakra in my logo.

(Sacred geometry is the study of different 3D shapes that have a sacred meaning in religion or spirituality; these shapes can be seen in places of worship, such as temples and churches.)

Stellar gateway: To strengthen your connection to other realms, the stars and other planets, use all the crystals that I have recommended above for the causal and soul star chakras. Additionally, you can use stellar beam calcite; this is known as the *crystal of ascension*, and brings peace, love, high vibrational energy and spiritual growth. I worked with this energy a lot during my training to became an angelic reiki practitioner, and still do now.

» Metatron's cube

This sacred geometric shape is associated with Archangel Metatron – the angel who helps with the creation of life, wisdom and connection with the divine light.

Metatron's cube is made up of 13 spheres, which connect together by lines forming other sacred geometric shapes like the ones discussed above. This is interesting because chakra 13 in my logo is associated with the stellar gateway, connecting to all divine energies of light in this planet and beyond. The cube represents the 13 chakras, and the different shapes formed represent the elements: water, air, wind, fire and ether. The cube explains the birth and the field of creation in all dimensions of time and space.

Crystal healing

The use of crystals in healing has been practised in many ancient civilisations, including the Egyptians, the Mayans, the Greeks, the Babylonians, the Chinese, the Indians and the Romans. For example, in ancient Greece, soldiers would rub hematite crystal all over their body for protection because of its high iron content; the Greeks associated iron with Mars, the Greek god of war. Hematite is an iron ore crystal. It has magnificent healing properties, and can help restore joy and vitality by absorbing any toxic energies trying to attach to your energetic auric field. Hematite is a key crystal to use for grounding, and enhances our connection to the earth.

Crystal healing is a non-invasive modality that cooperates with the inner healer in the body. It is a repair mechanism, and can achieve homeostasis (the balance of the body's energetic systems). Crystals are high-vibration; they work by connecting to our electromagnetic field and help to restore peace and relaxation, and to prevent dis-ease (this is when the body is stressed and is not at ease, leading to illness).

Crystals can help to uncover the psychosomatic causes of disease. The word 'psychosomatic' refers to emotions and mental attributes affecting the physical body. Crystals can help you in many ways, by improving your physical, mental, emotional and spiritual well-being. They can help remove symptoms of certain diseases, facilitate life choices to promote better health, and speed up changes in your life.

Having a healing session with an experienced crystal healer could help unblock the chakra or chakras by balancing the energy flow in the body. The crystal used can help in the healing process. To enhance this treatment, a crystal grid of four clear quartz crystals can be used around the crystal for the specific chakra, and the healer uses the master clear quartz in their hand to remove energy (anticlockwise rotating) or increase energy (clockwise rotating) once all the grids have been placed around the chakras. The recipient gets a much more relaxing treatment when the clear quartz grids are in place. The clear quartz amplifies the energy, and expands the auric field of the recipient so that more healing is achieved. The crystals and grids are left in place for about 20 minutes; this is so the recipient gets maximum benefit and relaxation while the crystals do the healing.

If you are interested in becoming a crystal healer, I recommend taking a course with Philip Permutt, a crystal healing expert. It will be a long journey, as there is so much to learn – but, by the end of it, you will be able to use specific crystals to relieve symptoms of different illnesses.

Working with a pendulum

A pendulum can help with your healing work, and help you to choose the right crystal or crystals you need at a given time.

- You first need to buy a pendulum from a crystal shop and cleanse it as described earlier.
- Hold the pendulum from the chain, and establish which way the pendulum moves for a yes and a no answer. For example, when I work with my pendulum and I ask it a question such as 'Do I need this rose quartz crystal?': if the answer is yes, my pendulum will move from side to side; if it is no, it will spin round in a clockwise direction.
- I use my pendulum when I am giving a crystal healing session. Once my client is lying down, I get the pendulum, go through each of the seven chakras and ask, 'Is this chakra balanced?'
- I make a note of this for all seven chakras. I then place the seven

chakra stones described above on the individual chakras and ask again, 'Is this chakra balanced?' I record the answers.
- I leave the crystals to do their magic for 15 minutes.
- I ask again, for each of the chakras, 'Is this chakra balanced?' If any are out of balance, I form a clear quartz grid, which amplifies the healing as described above.
- I continue to do this until all the chakras are fully balanced.
- This is known as *dowsing* and is the most ancient form of divination. It is thought that the pendulum is moved by the energy of your soul/spirit and your inner knowing.
- Dowsing is commonly used to locate groundwater, buried metals or crystals, gemstones, oils and many other things without the use of scientific apparatus.

Strengthening your aura

There are specific crystals you can use to strengthen your aura, such as labradorite, red jasper and smoky quartz.

Labradorite is a master healer crystal for the spirit; it deflects unwanted energy and negative thoughts. It stabilises the aura and enhances the flow of energy in the chakras. It unites logic and eliminates mental confusion. It is a wonderful protection crystal to work with. It is a magical crystal that changes colour depending on the reflection of the light. It would be a good idea to have this crystal in your pocket when you are out in public, as it has the ability to filter energies; this indicates that it will only let positive energy into your aura.

Red jasper is a very nurturing crystal – it supports and promotes tranquillity, and provides protection from unwanted negative energies around you that try to enter your auric field. It promotes the yin and yang energies in the body for complete balance. It cleanses the aura and aligns the chakras to make it easier for your spirit to manifest on earth.

Smoky quartz is used to remove negative energies and transmute them into positive ones. It removes environmental pollution to cleanse your aura. This chakra is great for grounding, as it has a great affinity with the earth element, and helps in times of stress, depression and anxiety by bringing you harmony and helping you release any energies that are no longer serving you. This crystal also helps to absorb any physical pain you may be experiencing as a result of absorbing negative energies.

Clear quartz is used for cleansing, protecting and strengthening your aura. This crystal is the master healer crystal for the mind, body and soul; it contains the full electromagnetic spectrum of visible light (please refer back to Chapter 1). This means it contains all the colours of the rainbow within it, and has different vibrational frequencies arranged by colour – which is why it is known as the *master healing crystal*: it is the most powerful amplifier of healing energy.

My crystal box

I have a selection of crystals I like to work with to enhance my experience and work. I have these stored in my magical crystal box. The main reason I use this set of crystals is that I have received huge benefits for my mind, body and spirit. I also have a large selection of crystals that I use during meditation to connect with other spiritual realms.

For protection, I use black tourmaline, a selenite wand and a celestine sphere.

For healing myself, I use my large clear quartz pointed crystal (which has been attuned to my energy, and was given to me during my angelic reiki training to become a practitioner), and rose quartz and green crystals including aventurine (I have many). I have a green amazonite on a chain and wear it just above my heart – it is extremely soothing and very therapeutic when I need some healing. I use green fluorite when I have emotional upsets.

For connecting to the angelic realm, during meditation I hold an angelite crystal. I have a Lemurian crystal attuned to angelic healing energies, which I really love to use; and an angel aura quartz, which enhances my connection to the angels.

For connecting to mermaids, dolphins and Atlantis, during meditation I hold my larimar crystal; and I have many blue crystals, including kyanite, aquamarine, blue lace agate, lapis lazuli and aqua aura that I work with and have on my mermaid altar.

For connecting to my clairvoyance and intuition, I use amethyst and tiger's eye.

My favourite crystal is labradorite. I love working with this crystal – it is so magical, it has different flashes of light, and it helps me to balance my spiritual thinking with my scientific thinking and with scientific analysis. It helps to balance my chakras and my auric field. I take this crystal everywhere I go. Along with labradorite, I use rainbow fluorite, as this helps with my thoughts and knowledge so I can express myself. It connects my throat (communication centre) and heart (love centre) chakras with my third-eye (intuition centre) and crown (mind centre) chakras.

Dalmatian stone is another great crystal I carry with me, as it repels negative energy and protects my auric field, helps balance yin/yang energy and helps balance my spiritual mind and my scientific mind. This is great for when I am learning, and helps me to achieve my goals.

Your crystal experience

There are so many different crystals out there, and you will choose your own crystals that you are drawn to. You need to feel into the crystal and see which colour, size and shape resonates with you. The crystals I have explained in this chapter have generic properties. For your personal experience, you may experience different sensations and feel different benefits. For example, clear quartz crystal is used to connect with the crown and causal chakras to enhance your connection with consciousness and spirit; people have reported that this crystal makes them feel grounded and would use it to balance their root chakra.

If you are just beginning your journey with the crystals, I highly recommend the book *The Crystal Experience* by Judy Hall. It has an accompanying CD, and many written exercises to help you use your crystals for specific purposes. Judy Hall is considered a crystal expert in this field, and has written numerous books on crystals (please see the reading section for a list of books that I recommend).

Activity: *Connecting to your crystal*
You can connect to your crystal through a visualisation meditation, as follows:

- Hold your crystal in your hand.
- Sit down or lie down; make sure you are nice and comfortable without any disturbances.
- Close your eyes, and imagine you are the colour of your crystal and everywhere you look is this colour.
- You are in a beautiful meadow; it is springtime.
- You walk towards this beautiful waterfall; behind the waterfall, there is a cave.
- You enter the cave and, to your surprise, it is made out of your crystal.
- You hold your crystal to your heart and ask it if there is anything you need to know at this time in relation to your physical, emotional, spiritual and mental health.
- Spend some time there connecting with your crystal.
- When you are ready, open your eyes.
- Look at your crystal. Does it look different now? Do you feel more connected to it?

This meditation is a nice way to work with your crystal, and you can do this every time you are working with a new crystal, or even with your existing ones. Use Table 6.3 to record the experiences you have with your crystals:

Table 6.3: Experiences with your crystals

Crystal	Colour	Shape	How does it make you feel before meditation?	How does it make you feel after meditation?

Activity: *Connecting to all 13 chakras with crystal meditation*

- Find a nice, quiet place to lie down.
- Place 13 crystals (one for each chakra) on your body or alongside it.
- Close your eyes.
- Connect with each chakra and its accompanying crystal, starting with chakra 1 – root chakra, and going all the way up to chakra 13 – stellar gateway.
- For each chakra, say in your mind, 'I want to connect to you, (name of chakra), beautiful vortex of energy, with this crystal (name of crystal)'.
- Repeat this sequence twice, and take two deep breaths.
- Once you have reached chakra 13, activate your pillar of light by saying these words out loud: 'I activate my divine light and connect to Metatron's cube, the symbol of life!'
- Visualise bright white light beaming from the top of your crown chakra and extending outwards as far as possible.
- Relax in this crystalline energy for 15 minutes.
- When you are ready, open your eyes.
- Repeat this meditation as much as you need to.

This brings us to the end of Chapter 6. Next, you will explore another type of gift from nature – essential oils made from flowers, plants, trees and other earthly substances – and how you can use them in aromatherapy.

Reading

Hall J. *Crystal Healing*. Octopus Publishing Group, 2011.

Hall J. *The Crystal Bible*, Volumes 1–3 book set collection. Octopus Books, 2009–2012.

Hall J. *The Crystal Experience*. Pyramid, 2014.

Moorey T. *Crystals Made Simple*. Pyramid, 2017.

Permutt P. *The Crystal Healer: Crystal Prescriptions That Will Change Your Life Forever*. CICO Books, 2007.

Permutt P. *The Little Pocket Book of Crystal Healing: Crystal Prescriptions That Will Change Your Life Forever*. CICO Books, 2017.

Raphaell K. *The Crystalline Transmission: A Synthesis of Light*. Aurora Press, 1990.

CHAPTER 7

Essential oils and aromatherapy

Essentials oils have been used for their medicinal properties and in fragrances for 5,000 years. In ancient Egypt, essential oils such as frankincense and cedar were used to heal wounds and prevent inflammation. The Egyptians would cover mummies in myrrh to protect the body and help the soul reach the afterlife. The ancient Greek physician Hippocrates, who was known as the father of medicine, used herbs to cure illness and disease. He discovered 300 different types.

Nowadays, essential oils are used in most European countries in complementary therapies such as homeopathy, and in aromatherapy to treat specific medical issues such as depression, body aches and pains, and dry skin.

What are essential oils?

Essential oils are pure, natural products that cannot be synthetically imitated. This is because of their active and specific ingredients. Essential oils can be used as a therapeutic healing product and as fragrances, whereas artificial oils (which are artificially made in a lab and will contain other chemicals/toxins) can only be used as fragrances. These fragrances are perfumes that you can buy in a shop.

Essential oils are nature's gift to us all. They are made from flowers, plants, barks of trees, and resins from the earth. The oils can be used in a concentrated form, or can be bought in incense sticks or in beauty products such as creams, hair products, bath salts, oils and bath foam. Different oils are used for different purposes and are beneficial to the mind, the body and the soul. Oils can be extracted from the flower and bark by different methods, which include the following.

» Steam distillation

This is the most popular method of extracting the essential oil from plants, roots, bark and resins. The plant material is placed in a container, which is then placed in a beaker of boiling water. The steam that is produced contains the essential oil formed from the plant material; this goes into a condenser attached to cold water, to reduce the temperature of the mixture. This is to produce the floral water and the essential oil separately. The essential oil floats at the top of the water, and is ready to be collected and bottled. The floral water is collected and is used in cosmetic products. Figure 7.1 shows how an essential oil is extracted by steam distillation.

» Cold pressed extraction

This method is mainly used for citrus fruits such as oranges, limes, mandarins, grapefruit, lemons and bergamot. I love working with these oils because they are so refreshing, energising and uplifting. The cold press method uses the peel of the fruit to extract the essential oil, and mechanically presses the fruit down to extract the liquid. The oil layer lies on the top of the liquid, and is removed to be bottled up as an essential oil. To make pure essential oil, you will need a lot of fruit peel!

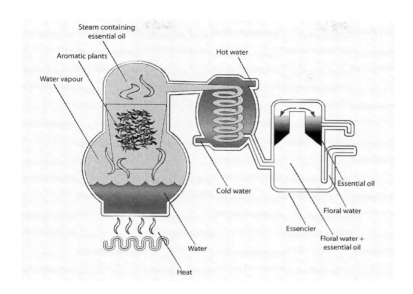

Figure 7.1: Steam distillation

» Absolute extraction

This method is used when flowers are too delicate to withstand the extensive heat environment in steam distillation. The flowers are blended with a solvent, and, after about a month, the essential oil is produced and removed. This product is called an *absolute* because it contains some alcohol. These oils tend to be more costly because of the number of flowers needed and the time taken to produce the essential oil. An example of this is jasmine essential oil.

» CO_2 extraction

This method uses carbon dioxide (CO_2) instead of heat. It involves putting the CO_2 under high pressure to convert it from a gas form into a liquid form, which then acts as a solvent. The solvent diffuses through the plant material and produces the essential oil. The only issue with this method is that it produces additional constituents, such as solvents; as it is a fairly new method, more experimentation is needed so that it can be used more routinely.

Why use essential oils?

The main reason for using essential oils is that they can help raise your frequency – this means they can benefit you in a variety of ways depending on the oil you use. Each essential oil has specific characteristics and works with different body systems (please refer back to Chapter 3 to remind yourself of these systems). Essential oils can help and heal all areas of your life. This is because they hold their own energetic frequency, as we do.

Plants, trees and flowers are all living beings, and they too each have their own frequency. Trees communicate with each other via their root system and have the ability to adapt to the different seasons. For example, during autumn, trees lose their leaves, which change colour from green to red, yellow, brown and orange. Is this not amazing? This is why we, as human beings, should show respect and gratitude to these plants and trees: actually, without trees producing oxygen and cleaning pollution in the air, we would not be alive. Next time you are out in nature, go to a tree, hug it and say, 'I am grateful for the life you bring to the animals, to the earth and to me!' Send love and healing to the tree.

Safety tips when using essential oils

- Store oils in a dry place away from direct sunlight.
- Make sure the essential oils you buy are in amber or blue bottles.
- Keep them away from children and pets.
- Keep them away from naked flames.
- Make sure lids are tightly closed after use.
- If you are pregnant or breastfeeding, seek medical advice from your doctor or a certified aromatherapist.
- Never use undiluted oils on babies, children and animals.

- If you wish to take any essential oils internally, seek professional medical advice from a qualified homeopathic doctor.
- Some oils are photosensitive; if you have applied these on your skin and you go out in the sun, it will cause irritation and make you feel unwell.
- Always check the instruction booklet of every essential oil you use; if the oil you have bought does not have one, personally, I would not use it.
- Some oils can irritate the skin in their neat concentration; check the oil's safety leaflet.
- If you want to use the oil in a steam inhalation, always use hot water, but not boiling water.
- If you use an oil and it starts to burn your skin, wash thoroughly under cool water.
- If the oil gets into your eyes – for example, if you have used it in a bath at a high concentration – you must wash your eyes with lots of water.

How do essential oils work?

Essential oils help to bring peace and balance to us – mentally, physically, spiritually and emotionally. Here is how they can achieve this:

- **Mind**: They work on the mind by stimulating focus, through relaxation, and through balancing emotions by giving calming and uplifting effects. Examples of essential oils to aid the mind include chamomile, lavender, rosemary, lime, lemon and grapefruit.
- **Body**: They work on the body to relieve pain, reduce inflammation, work as an antiseptic, provide detoxification and cleansing, regulate hormonal imbalances and boost immunity. Examples of essential oils to aid the body include jasmine, sage, tea tree, orange and basil.

- **Spirit**: They work by promoting mindfulness; aiding spiritual practices such as prayer, meditation and balance; and grounding and protecting the energetic aura and chakra system. Examples of essential oils to aid the spirit include myrrh, frankincense and sandalwood.

My top 20 essential oils

There are so many different essential oils – over 200. If you want to learn more about all of them, please refer to the reading list at the end of this chapter. I will take you through the top 20 essential oils that I use regularly. I will describe their characteristics and how to use them in Table 7.1.

There are many herbs that you can use in cooking that will be beneficial to your health. I recommend using organic herbs that are free of pesticides, so you will experience their full medicinal benefits. If you have a garden or an allotment, you can grow these yourself!

Below are the main herbs you can use:
- **basil** – fights ageing, treats respiratory system and helps digestion
- **dill** – treats heartburn, colic and gas
- **parsley** – anti-inflammatory, controls blood pressure
- **oregano** – soothes stomach muscles
- **mint** – helps with upset stomach, refreshing if drunk as a tea, alleviates nausea
- **clove** – antimicrobial
- **rosemary** – antioxidant, aids digestion
- **sage** – antiseptic and antibiotic
- **fenugreek** – flushes out toxins
- **black pepper** – helps relieve indigestion
- **cinnamon** – helps lower blood pressure, increases metabolism to aid weight loss
- **turmeric** – anti-cancer, helps boost immune system
- **thyme** – antioxidant, anti-inflammatory, antibacterial.

Essential oil	Latin name	Extraction method	Characteristics	Region found	Uses	How it works
Rose	Rosa damascena	Steam-distilled to produce a yellow or amber oil with an intense floral aroma	Calming, uplifting and rejuvenating	Most countries – a commonly grown garden shrub	Skin care, diffuser, massage oil	Antidepressant, repairs and heals skin, enhances well-being, supports women's health, antibacterial
Jasmine	Jasminum grandiflorum	Solvent extraction to produce a thick orange oil, known as an absolute, with a floral aroma	Uplifting and relaxing	India, Iran	Fragrance, skin care, diffuser	Soothes and tones skin, enhances well-being, fights the cold, aids depression and helps with libido in both men and women
Orange	Citrus sinensis	Cold pressed to produce a pale orange oil with a fresh fruit aroma	Uplifting, detoxifying, calming and rejuvenating	China, Cyprus	Bath oil, massage oil, diffuser	Aids detoxification, revitalises and brightens skin, improves digestion, relieves anxiety, boosts immunity, appetite stimulant, antidepressant
Frankincense	Boswellia carterii	Steam-distilled from tree resin to produce a yellow oil with a sweet and spicy aroma	Calming, soothing and invigorating	Eastern Africa and Oman	Ointment, diffuser, massage oil, facial skin cream	Anti-inflammatory properties, stimulates cell regeneration, relieves anxiety, wound healing, treats respiratory illnesses such as asthma, antiseptic, fights colds and flu
Lime	Citrus aurantifolia	Cold pressed to produce a green oil with a citrus scent	Refreshing, energising and cleansing	Sri Lanka	Diffuser, skin care, bath oil	Aids detox, balances oily skin and hair, stress relief, antiseptic, stimulant

Essential oil	Latin name	Extraction method	Characteristics	Region found	Uses	How it works
Mandarin	*Citrus reticulata*	Cold pressed to produce an orange oil with a citrus scent	Cheerful, energising and uplifting	China and Far East	Diffuser, massage oil	Antiseptic and antifungal, tones skin, detox, aids digestion, safe to use in aromatherapy for children, mood lifter, relieves stress and anxiety
Peppermint	*Mentha piperita*	Steam-distilled to produce a yellow oil with a sweet smell	Stimulating, soothing and refreshing	Worldwide	Steam inhalation, massage oil, diffuser	Effective deodorant, helps with digestive upsets and nausea, relieves headaches, soothes insect bites
Lavender	*Lavandula angustifolia*	Steam-distilled to produce a yellow oil with a strong floral aroma	Calming, rejuvenating, soothing and relaxing	UK and France	Diffuser, bath oil, massage oil, perfume	Heals skin, promotes restful sleep, helps with anxiety and depression, acts as antiseptic
Ylang-Ylang	*Cananga odorata*	Steam distillation or water distillation to produce a yellow oil with an oriental floral aroma	Uplifting and relaxing	Madagascar, Philippines	Ointment, diffuser, massage oil, facial skin cream	Anti-inflammatory, calming, acts like a sedative, eases anxiety, treats depression, treats sexual problems, lowers blood pressure and heart rate
Lemon	*Citrus limonum*	Cold pressed extraction to produce a yellow oil with a citric scent	Invigorating, cleansing, uplifting and detoxifying	Native to India; arrived in Europe in 12th century	Massage oil and in a diffuser	Tones skin, acts as an antiseptic, aids detox, eases digestion, lifts mood by focusing on the mind, increases alertness

Essential oil	Latin name	Extraction method	Characteristics	Region found	Uses	How it works
Tuberose	*Polianthes tuberosa*	Solvent-extracted to form an orange-brown viscous oil with a spicy, sweet aroma	Calming, uplifting	France	Diffuser, massage oil	Balances emotions, antidepressant, provides relief from nervousness, anger, exhaustion and stress, mild sedative, combats body odour, boosts circulation, acts as an aphrodisiac
Cedarwood	*Cedrus atlantica*	Steam distillation of tree bark produces a brown oil with a woody scent	Uplifting and soothing	Lebanon	Ointment, massage oil, hair tonic, diffuser	Treats depression and stress, soothes and heals skin, pain relief, antibacterial and anti-inflammatory, detoxification, fights colds, lifts mood
Cinnamon	*Cinnamomum zeylanicum*	Steam distillation of inner bark and leaves to produce a brown oil with a warm, spicy aroma	Energising and stimulating	Sri Lanka	Foot spa, diffuser	Soothes aches, aids digestion, balances emotions
Neroli	*Citrus aurantium*	Water-distilled to produce a yellow oil with a sweet, floral scent	Calming and refreshing	China	Diffuser, perfume, skin care	Relieves stress, improves digestion, promotes healing, helps with both depression and anxiety
Tea tree	*Melaleuca alternifolia*	Steam-distilled to produce a clear oil with a eucalyptus aroma	Soothing, medicinal and deodorising	Australia, New Zealand	First aid, foot spa	Boosts immunity, fights infection, dental care, heals cuts and bruises, fights body odour, fights colds and flu, controls acne

Essential oil	Latin name	Extraction method	Characteristics	Region found	Uses	How it works
Grapefruit	*Citrus paradisi*	Cold pressed from peel of fruit to produce a yellow oil with a sweet aroma	Energising and uplifting	Caribbean	Massage oil, fragrance, bath oil	Combats tiredness, soothes aches, aids detox, tones skin, helps with stress, anxiety and depression
Lemongrass	*Cymbopogon citratus*	Steam distillation of chopped grass to produce an amber oil with a fresh aroma	Refreshing, stimulating, deodorising	South-east Asia	Skin care, diffuser, first aid	Antiseptic and antifungal, insect repellent, enhances well-being, aids digestion
Roman chamomile	*Anthemis nobilis*	Steam-distilled to produce a light blue oil with a sweet aroma	Soothing and calming	UK	Bath oil, massage oil, skin care	Antidepressant, relieves PMS symptoms, soothes digestive system, anti-inflammatory
Clary sage	*Salvia sclarea*	Steam-distilled to produce a pale green oil with a sweet, nutty aroma	Uplifting and reviving	Southern Europe	Bath oil and massage oil	Lifts mood, relieves aches and pains, aids women's health, boosts circulation, aids digestion
Sandalwood	*Santalum album*	Steam-distilled to produce a pale gold oil with a sweet, woody aroma	Healing and soothing	Southern India	Diffuser, used in aftershave balm	Restores health, vitality and well-being, eases breathing, great antiseptic, helps with sore and dry throats, used for spiritual protection, aphrodisiac for men

Carrier oils

Most essential oils must be diluted before you use them on the skin, face and hair – you can do this using carrier oils. The most common carrier oils are as follows:

- **Organic sweet almond oil**: This is the most versatile, and I use it as a carrier oil for all my essential oil blends. It is suitable to use on babies, infants and animals.
- **Organic coconut oil**: This is extremely moisturising and can be used when making hand creams, shampoo and conditioner for all hair types. If you use it neat without essential oils, it is good for protecting gums and teeth and can be used as a make-up remover.
- **Organic olive oil**: Contains antioxidants and fatty acids, so would be good to use as a carrier oil with most essential oils. It can used in making skin creams and hair products.
- **Organic argan oil**: Contains high levels of antioxidant vitamin E, so would be useful to mix with essential oils used for moisturising the skin, for hair and hand conditioner. Helps protect against environmental pollution.

What is aromatherapy?

Aromatherapy is the technique that uses essential oils to enhance well-being – psychologically, physically and emotionally – to bring harmony, clarity and balance to your life! This technique can help the body by assisting with hormone imbalances, low immunity, skin conditions, body aches, asthma and other respiratory disorders, and can help the mind with disorders such as depression and anxiety.

» History of aromatherapy

The most powerful of our five senses is scent. It is the most sensitive: certain scents we love, and others we hate. This is due to the association we have with certain smells. Smells are known to activate specific memories, emotions and feelings. Scent must have guided our ancestors to choose herbs and food. In the Stone Age, burning different types of wood led to the stimulating effects of sage and rosemary in the air. The ancient priestesses in Egypt would burn resin from frankincense trees to clear their minds, as well as using frankincense in rituals and ceremonies. The Romans used oils for massage and used herbs in their homes. Orange and cloves were used to ward off the Black Death in 17th-century England.

Rene-Maurice Gattefosse, a French chemist, was regarded as the father of aromatherapy, and actually invented the word 'aromatherapy' in 1937. It all began when, one day, he was working in his chemistry laboratory and discovered the therapeutic effects of lavender oil. He burned his hand and placed it in a bowl of oil, not knowing that it was lavender oil. To his surprise, later that evening, his hand was healed and there were no signs of any scarring. He was so excited about discovering the healing benefits of lavender that he began his research and started working with other essential oils such as lemon, thyme and clove. He used these oils on injured soldiers in military hospitals to treat their wounds, as they had antiseptic qualities. He found that using these oils also speeded up the healing process. He also went on to develop distillation techniques to extract the oils, and had a farm where he grew many aromatic plants such as lavender, so he could supply pure essential oils to other people.

How does aromatherapy work?

The tiny molecules of the essential oils enter the body as you breathe them in. They are carried by the olfactory nerve cells in the nose. Once they enter the blood stream, a message is sent to the brain, which associates the scent with a specific receptor in the limbic system. The limbic system controls our emotions, behaviour, passion and memory. The message produces an emotional response and releases neurotransmitters such as serotonin, which responds by making you feel happy and relaxed.

This can also help the body in a physical way. For example, people who suffer from depression tend to have fatigue because of not sleeping well. An oil that is commonly used to alleviate the symptoms of depression is lavender. This essential oil helps with sleep, brings calmness to the body and, in time, brings more happiness to the person using it. Essential oils can be used in a bath and be put into a diffuser; the essential oil is placed in water, and cold steam from the diffuser carries the oil molecules and fills the air. You can use this at home, in your car or at work. For example, if you are working as a reiki healer, you could incorporate using a diffuser with a specific essential oil with the healing session you are giving to enhance the treatment.

Why is it so important to incorporate essential oils and aromatherapy in your everyday life?

The main reason is to bring balance, relaxation and positive energy back to your body. Everyday life can be stressful and overwhelming, and we can come in contact with environmental pollution from cars, trains and other people. This can have a negative impact in an emotional way, making you feel exhausted; and in a physical way, causing damage to your body (cells, tissues, organs and body systems).

Oxidative damage occurs when there is an excess of free radicals in the body's cells absorbed from your environment. These can damage cells, proteins and DNA, which can contribute to ageing, cancer, heart disease and even premature death. Antioxidants in essential oils can help control free radical damage at the cellular level, supporting a healthy body and mind. Essential oils have extremely high levels of antioxidant capacity, and help reduce or prevent oxidative damage to the body.

Energy, frequency and vibration

If you remember, these concepts were discussed in Chapter 1 on energy. I want to remind you again of these terms, as they are very relevant here.

Energy is a type of force that cannot be created or destroyed, only transformed into another form; frequency is the number of waves that pass a fixed place in a given amount of time; and vibration means back-and-forth movement about a point of equilibrium. Every single living being is made up of energy. Negative thoughts can lower our frequency, whereas positive thoughts raise our vibration and bring us feelings of happiness.

This brings me to this famous quote by Albert Einstein:

> *Everything is energy and that's all there is to it. Match the frequency of the reality you want and you cannot help but get that reality. It can be no other way. This is not philosophy. This is physics.*
>
> *We are mass energy. Everything is energy. EVERYTHING!*
>
> Albert Einstein

Another famous physicist, Nikola Tesla, thought along the same lines as Einstein and produced this quote:

> *If you want to find the secret of the universe, think in terms of energy, frequency and vibration.*
>
> Nikola Tesla

Tesla was best known for his contributions to the alternating current (AC) electric system and the creation of the Tesla coil, which is still used in radio technology and is the foundation for wireless technologies that we use nowadays on a daily basis. Now can you understand the concept of energy and frequency.

Frequencies of the human body

The human body has its own frequency. Frequency is measured in megahertz (MHz). The frequency of a healthy human body has been suggested to be in the range of 62MHz to 70MHz. Our frequency is found in our body and auric field; energy disturbances will firstly show up in our aura, and will manifest as a disease or illness in the physical body. If the body's frequency goes below 62MHz, this indicates some sort of problem in the body: for example, a frequency of 58MHz indicates a cold or flu, 42MHz indicates cancer somewhere in the body, and 20MHz indicates the body is heading towards death.

Royal Raymond Rife invented a machine where low-level electromagnetic energies could be given to patients suffering with cancer. His work suggested that these energies were enough to change the frequency of a cancer cell and help relieve symptoms. These are still used as part of cancer treatment in clinics (please refer to the reading list).

Essential oils have a much higher frequency than the human body – for example, rose essential oil has a frequency of 320MHz, and frankincense has a frequency of 147MHz – which is why essential oils can increase the frequency of the body when used in aromatherapy.

Essential oils such as lavender, peppermint, tea tree and jasmine have been used to relieve side effects in cancer treatments such as chemotherapy, radiotherapy and hormone-specific treatments. Studies have shown that symptoms such as nausea, anxiety, depression and lowered immunity have improved when aromatherapy was incorporated with conventional medication.

Grade A therapeutic essential oils are the best for internal use, as they are pure and natural. These can be used for specific purposes in the body to raise the frequency and get us feeling and functioning normal again. This is achieved because the frequency of the chosen essential oil resonates with the frequency in the part of the body that is out of balance. I work with essential oils daily, and the ones I recommend most strongly are from Neal's Yard Remedies, as these are extracted from reliable organic sources: plants, flowers and resins.

What are the different uses of aromatherapy?

Aromatherapy can be used in many ways:
- therapeutic massage
- beauty products such as creams, oils and soaps
- sprays
- in a diffuser
- in a foot spa
- in a bath.

Massage

An aromatherapy massage is when a trained aromatherapist uses a blend of essential oils in a carrier oil and gently massages the skin. This could be on the hands, the feet, the face or even the whole body. The therapist rubs the oil blend, which is usually warm, on their hands, and then starts massaging pressure points in the body to relieve the pain. A few examples of oil blends are:
- **for the face** – jasmine and rose: great for the skin, and these oils are found in many face creams
- **for the body** – lavender and sandalwood: great for relaxation and grounding energy
- **for body aches and pains** – lavender and chamomile: great for relaxation and treating inflammation.

An aromatherapy massage can last up to 90 minutes. Please remember to go to a professionally trained aromatherapist with qualifications.

To find your local trained aromatherapist, go to the International Federation of Aromatherapists' website: ifaroma.org/en_GB/home.

Benefits of aromatherapy massage

- Relaxing.
- Soothing.
- Reduces stress levels.
- Protects and can stimulate the body's own natural healing mechanism.
- Relieves tension in muscles and joints.
- Helps you connect with natural, beautiful-scented aromas.
- Helps you feel balanced – mentally, emotionally and physically.

Essential oils versus pharmaceutical drugs

Essential oils are natural and grown organically, and contain hundreds of molecules. No two batches are ever the same. They restore natural function, have no adverse interactions, are antiviral, improve intercellular communication, correct and restore DNA damage, cleanse receptor sites, build the immune system, are emotionally balancing, work with the body's natural healing mechanism to restore balance, and lead the body towards independence and well-being. Most importantly, any side effects are actually beneficial and help other areas in the body.

Pharmaceutical drugs are unnatural and synthetically made in a laboratory, and contain one or two active ingredients. They inhibit natural function, disrupt intercellular interactions, block receptor sites, depress the immune system and are emotionally unbalancing. They have harmful side effects, and lead to disease and dependence.

Which one would you prefer to use to help your illness or disease? I know what I would do, and I now use essential oils daily. I hardly ever take medication unless I am in severe pain; I may take two painkillers. I will not take any other forms of medication or antibiotics. I use natural methods to heal and feel better: things such as lemon tea, chicken soup and honey, and the use of essential oils, always help to speed up my recovery.

Making your own aromatherapy products

As you start to work with essential oils regularly to help yourself and others, you will start to learn how to make your own products instead of buying them already prepared. I will share with you a few recipes so you can begin to do this.

» Making a clearing spray

Clearing sprays can be used to clear the energy of a room after you have given someone a treatment, and also if you need to change the frequency of the energy in a room.

Equipment and ingredients
- Amber glass spray bottle with a glass bead (100ml)
- Measuring beaker
- 100ml of distilled water (this is recommended because normal tap water may have bacteria in it; it also helps the product last longer)
- Organic rose essential oil
- Organic jasmine essential oil
- Organic frankincense essential oil

Method
- Using the measuring beaker, pour 90ml of distilled water into the amber glass bottle.
- Add 20 drops of each essential oil into the water.
- Add the remainder of the water.

- Put the lid on the bottle, and rotate it to make sure all the ingredients are mixed.
- Then spray and enjoy!
- Keep for up to 3 weeks.

» Making a moisturising cream for the face

This will soothe and regenerate skin cells.

Equipment and ingredients
- Glass stirrer
- A spoon
- Amber glass jar with lid (100ml)
- Organic coconut oil
- Organic rose essential oil
- Organic lavender essential oil

Method
- Fill the amber glass jar with coconut oil until it covers the top of the jar.
- Clean around the jar.
- Add 40 drops of lavender oil and 40 drops of rose oil.
- Stir mixture with glass stirrer.
- Put the lid on firmly, label and date.
- Store at room temperature; or, if the weather is warm, store in the fridge and take out 10 minutes before you want to use it.

Rose oil is extremely beneficial to use on the skin as it helps regenerate healthy skin, and lavender is extremely soothing and can help calm the skin.

» Making a cleaning spray for your bathroom

Sage has antiseptic properties and helps clear unwanted energy.

Equipment and ingredients
- Measuring beaker
- Amber glass spray bottle (500ml)
- Saucepan
- Hob
- Distilled water
- Sage leaves

Method
- Boil water in a saucepan, filled up to three-quarters; keep an eye on this, and top up water if needed during the boiling stage.
- Put sage leaves in the saucepan.
- Simmer for at least 4 hours or until you can see the leaves are starting to discolour and break.
- Drain the leaves and let the water cool.
- Measure 500ml of water with your measuring beaker, then transfer to the glass spray bottle.
- Label, date and store in a dry place. Use within a month.

The product you have made is a sage hydrosol; it is not the essential oil of sage, but still has the same antiseptic properties in it.

Activity: *Guided meditation to connect to your favourite essential oil*

- Find a nice, quiet place to begin your meditation.
- Get your favourite essential oil; rub some into the palms of your hands (don't use it neat – mix a few drops in an amber container with organic almond oil).
- Cup your hands and breathe in the beautiful aroma.
- Sit or lie down.
- Close your eyes and imagine you are the plant or tree that produces this magical essential oil.
- What does the plant or tree look like? Does it have a message for you?
- Spend some time connecting to the source of origin of your essential oil.
- When you are ready, open your eyes.
- Journal your experience.

Reading

Brown CW. *Sacred Oils: Your Guide to 50 Essential Oils and How to Use Them for Healing and Well-being*. New Burlington Press, 2019.

Curtis S, Thomas P, Johnson F. *Neal's Yard Remedies: Essential Oils*. Dorling Kindersley Limited, 2016.

Westwood, C. *Aromatherapy: A Guide for Home Use*. Amberwood Publishing Ltd, 1991.

Useful web pages

National Cancer Institute: the use of aromatherapy to combat side effects associated with cancer treatment – www.cancer.gov/about-cancer/treatment/cam/patient/aromatherapy-pdq

Essential Oils & More: essential oil application to use on your phone – https://play.google.com/store/apps/details?id=essential.oils.more&hl=en_GB

Energicx•USA: the frequency of the human body – www.energicxusa.com/frequency-of-human-body

National Cancer Institute: aromatherapy used to relieve symptoms of cancer drugs – www.cancer.gov/about-cancer/treatment/cam/patient/aromatherapy-pdq

Cancer Research UK: use of Rife machines in cancer treatment – www.cancerresearchuk.org/about-cancer/cancer-in-general/treatment/complementary-alternative-therapies/individual-therapies/rife-machine-and-cancer

Neal's Yard Remedies: organic essential oils – www.nealsyardremedies.com

CHAPTER 8

The brain and mindfulness

I begin this chapter with great excitement and joy – my work on this book is nearly complete, and this chapter is the one where I will be sharing all my knowledge, obtained from years of studies and experience in different areas related to this topic. I am like your big book of knowledge, and, for this chapter, I will unlock that book, and go into the depths of my mind, memory and knowledge to share it all with you.

I am sitting at my desk, it is dawn, the portrait of Albert Einstein is in my vision; I feel that if he was here right now, his message to me would be, 'I am so proud of you – you are nearly there – keep going and in time you will see the rewards.' I am sipping on my favourite tea: tie guan yin oolong, from Taiwan. I have sandalwood incense burning, as I find this opens up my mind and helps me to focus. At this present time I am feeling relaxed and peaceful!

Let me begin. For as long as I can remember, I have had a huge fascination in how people behave and their actions. At the age of 14, one of the subjects I chose to study at GCSE level was sociology: this social science is the study of the structure and function of society, including the relationships and behaviours of the people in that society. I really enjoyed this subject, and remember that my project was a study of antisocial behaviour. I was analysing how members of the public behaved in everyday situations and how they treated others. I recorded a wide spectrum of behaviours, ranging from extremely polite to extremely aggressive (the aggressive people scared me when I was as a teenager, and I would keep my distance).

Following on from this, I decided to do A Level Psychology at college, to go deeper into human behaviour and explore the mind. At first, I was unsure whether this would fit in with my other science subjects, Biology and Chemistry, but to my surprise they interlinked with each other very well, and I was certain I wanted to incorporate all three subjects in my higher educational studies. Psychology is the study of the mind – unconscious and conscious – how we feel and think, and understanding our thoughts, actions and behaviour. I learned so much on this course that I wanted to go even deeper and learn about the structure of the brain and how it functions and behaves in health and disease. I went on to pursue an MSc in Neuroscience at the Institute of Psychiatry at King's College, London. Neuroscience is the scientific study of the nervous system: its cells (neurons) and how they communicate with each other. It covered health and disease in both humans and animals. This degree was fascinating, but was the most difficult thing I have done in my life. I found it hard because I was working full-time and studying; it was a stressful time, but was rewarding and taught me discipline, structure, good timekeeping and excellent organisation skills.

Let us begin our journey now and explore the brain and the mind, beginning with the smallest part of the brain: the neuron.

The neuron

Neurons are the cells of the brain that carry impulses; these are electrical messages that move around the body. Each neuron consists of a cell body, which contains a nucleus (the control centre of the cell) and the cytoplasm, which contains *dendrites* and *axons*. Dendrites carry impulses away from the body, while axons carry them towards it. A neuron has a myelin sheath that surrounds the axons, to help with electrical conduction. A neuron communicates with another cell by connecting its dendrite to the axon of the other neuron, and an electrical transmission occurs. Information is passed by the synapse region, where the membranes of adjacent cells are in close opposition to one another.

Chemicals known as *neurotransmitters* in the synaptic vesicles are released from the terminal end of the axon of one neuron and bind to the receptor site on the terminal dendritic end of the other neuron to produce the response. Different neurotransmitters are released to produce specific responses: for example, serotonin is a neurotransmitter that, when released, causes a happy response in our mental state. This can occur when there is a stimulus: for example, when essential oils such as bergamot, lavender and lemon are inhaled, our sense of smell instructs the brain to release serotonin and dopamine neurotransmitters, which bring feelings of joy and calmness.

Imbalances in these neurotransmitters can affect your mood, sleep, memory and libido, and can cause addictions to drugs, alcohol and food. Serotonin contributes to our well-being and happiness, and dopamine contributes to pleasure, motivation and learning.

There are three different types of neurons:
- **Sensory**: Nerve cells that carry information from peripheral receptors to the central nervous system (CNS) and connect to our senses.
- **Intermediate**: These connect sensory and motor neurons.
- **Motor**: These carry impulses from the CNS to the effectors that connect to the muscles, resulting in movement of the limbs.

Figure 8.1 on the next page shows the structure of a neuron.

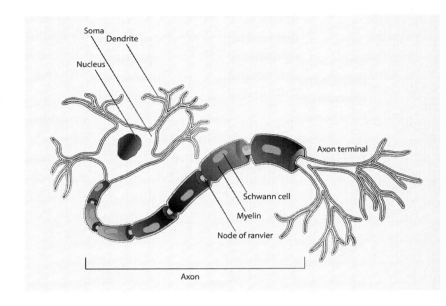

Figure 8.1: A neuron

The brain

The brain is probably the most complicated structure in the universe. It is the centre point of thought, intelligence, movement, learning, emotions, hormone regulation and so many more functions. The brain connects the mind to the body. Without the brain, the body will not function. In simple terms, the brain responds to a stimulus of some kind; this is then connected to a receptor in the brain, which produces a response. For example, if you pick up a hot cup of tea, the brain will instruct the arm to pick up the cup by the handle rather than the main body of the cup, so that you don't burn your hand. The brain is able to store memories, so if you have had a bad experience in a specific place, your brain reminds you not to go there again.

The brain is split into what I would call 'compartments', and different regions of the brain help the body to function – physically, mentally and emotionally. Different areas of the brain are used for receiving and sending out impulses that control specific functions. Figure 8.2 illustrates the structure of the brain and its different areas.

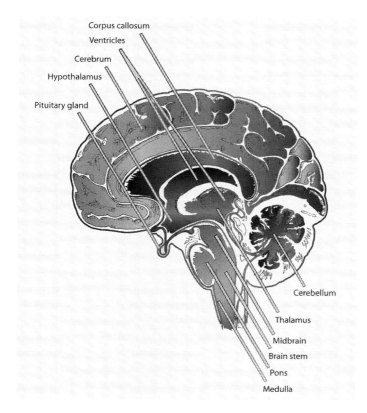

Figure 8.2: The human brain

The main areas of the brain are as follows:
- **The cerebrum** is responsible for decision making, and controls physical and mental activities. It is a large area of the brain and controls various voluntary actions. Sensation, reasoning, learning, memory, speech and language all take place in the cerebrum. *Voluntary actions* are ones we know we are doing, such as walking across the room: this will require conscious activities in the brain.

When the nerve impulses reach the brain, they are analysed before a response is decided upon.

- **The cerebellum** coordinates muscle movement and balance as instructed by the cerebrum, and is involved in the learning of motor skills.
- **The hypothalamus** controls all involuntary actions. *Involuntary actions* are ones that the brain has no conscious control over. We also call these *reflex actions*. An example is pulling your hand away from a boiling kettle. Your unconscious mind knows that there is danger ahead and responds by moving your hand away. This area of the brain controls homeostasis, such as body temperature and water potential, and works to bring the body back to equilibrium if it falls out of balance. It controls body temperature and uses physiological responses to maintain body temperature so it stays in the range from 36.1°C to 37.2°C. Mechanisms include sweating when there is hot weather, to release water molecules that will help to cool the body down; also, in cold weather, the hypothalamus instructs the body to shiver, as the resulting vibrations bring warmth.
- **The brain stem** connects the cerebrum with the spinal cord and is made up of the midbrain, the pons and the medulla.
- **The medulla** controls involuntary actions such as heart rate, digestion and breathing under the instruction of the hypothalamus.
- **The pons** conveys information about movement from the cerebrum to the cerebellum.
- **The midbrain** controls many sensory and motor functions, including eye movements and the coordination of visual and auditory reflexes.
- **The thalamus** relays impulses from all sensory systems to the cortex, which then sends messages back to the thalamus.
- **The corpus callosum** connects the left and right cerebral cortex. Its function is to integrate motor, sensory and cognitive performances between the two sides of the cerebrum.

- **Ventricles** produce cerebrospinal fluids, which are needed to provide mechanical and immunological protection to the brain.
- **The pituitary gland** controls the activity of most hormone-secreting glands.

» The nervous system

The nervous system is the most complex and sophisticated network known to man. Its main functions are to detect changes in the internal and external environment, and bring the right responses to the glands, the muscles and all the organs in the body. As we saw in Chapter 3, the nervous system is made up of two parts: the central nervous system (CNS) and the peripheral nervous system (PNS).

The CNS consists of the brain and the spinal cord; it is the most complicated area of the nervous system, and is where most nerve cell bodies are located and where most synaptic connections occur. The CNS receives and integrates sensory input to formulate motor output. The PNS consists of the receptors and effectors to the CNS outside the brain and spinal cord.

» The limbic system

The limbic system connects areas of the cerebrum in a complex network of neuronal cells in the hypothalamus, thalamus, hippocampus and amygdala, which govern higher mental functions such as emotions, learning, memory and behaviour. A memory is when we hold onto a thought of something that has happened to us in the past, ranging from a word to an emotional experience that has shaped our lives, and keep it locked in our brain until we think of it again. Learning takes place when we retain and make use of past memories.

» Neurodegenerative diseases

These diseases are incurable and affect the normal functioning of the brain. They are progressive and get worse over time, which eventually causes death in the sufferer. Normal nerve cells experience neuronal cell death, and will contain cells with abnormalities. It is the abnormalities in the cells that effect cognitive functions such as memory and speech, along with motor functions such as balance and movement.

The most common neurodegenerative disease is Alzheimer's disease (AD) and is known as a type of dementia. It targets the region of the brain associated with memory, behaviour and intelligence. Loss of memory begins with forgetfulness; as the disease progresses and becomes more advanced, the person will lose their memory of doing day-to-day tasks, and will not recognise their immediate family members such as their partners and children. There will also be a decline in intellectual activity and deterioration in behaviour. The disease is diagnosed by detecting neuronal and synaptic loss in the hippocampus and the cerebrum. Loss of acetylcholine neurones is another hallmark of AD.

Another common neurodegenerative disease is Parkinson's disease (PD). This is a type of ataxia – a disorder that affects coordination, balance and speech. PD is a movement disorder characterised by continuous tremors in the limbs, lack of sensory coordination and a tendency to feel exhausted. This condition is thought to be caused by the loss of neurotransmitter dopamine and the death of neuronal cells in the part of the brain called the *substantia nigra*.

The mind

The mind is interesting to define, as it has so many components to it. Part of it controls thoughts, emotions, behaviours, memories, learning, perceptions and your sensory consciousness, including the five senses: sight, hearing, smell, taste and touch. The other part of the mind involves the spiritual concepts such as imagination, philosophy, religion, sleep and meditation.

Sigmund Freud, the Austrian neurologist and founder of psychoanalysis (a method based on the belief that all people possess unconscious thoughts, feelings, desires and memories) created his own model of the mind. This model represented it as an iceberg floating in the sea. He believed the mind was split into three parts:
- **Conscious**: Having awareness of something and directing your focus on it in the present moment – the tip of the iceberg (10 per cent). It is the home of the ego and decision making.
- **Subconscious**: The storage area of any memory needed for a quick recall – the mid-part of the iceberg, floating above the sea (50–60 per cent).
- **Unconscious**: For storage of all our past experiences and long-term memories – the part of iceberg deep in the sea that you can't see (30–40 per cent). It is the home of the part of the psyche that contains repressed ideas.

I studied the work of Freud in great detail as part of my studies in A Level Psychology.

» What is mindfulness?

Mindfulness is being in the present moment and enjoying what life has to offer without being judgemental; and reconnecting with the simpler things in life. It is being at one with life, nature, the earth, the animals. It is activating all the senses that make you feel happy to be alive. Mindfulness is based on your own perception of it and what mindfulness is for you: this may be going for a walk in the park, doing some gardening or painting, or reading your favourite book – whatever it is, you must do it as often as you can. Mindfulness is also being in the here and now, and connecting healthily with the energies around you. There are so many books about this topic, and there are even colouring books that apparently can help you to be in this state of mind.

I will explain what mindfulness means to me. I like to go to a particular park close to my home. It is the most peaceful place in the world for me. It is my mindfulness. I connect to the sky and the clouds there, because the park is on a very steep hill. I activate my senses; I hear the wind blowing in my ears, the birds singing, see a magnificent view of London and smell all the aromas of the trees, the soil and the flowers. I feel at peace here, and I can connect to all four elements. I love being here no matter the weather – even in rainy weather, and even when it is a really busy summer day. I find my favourite tree, which I associate with the fairies, and sit under the tree with my back resting on the trunk. I close my eyes and sit there in silence, connecting to the energies all around me. Figure 8.3 shows my place of mindfulness.

Figure 8.3: My place of mindfulness

Is there a place that you like to go to experience mindfulness?

This very famous quote was written on a piece of paper by my famous friend Albert and sold for $1.3 million. This quote speaks out to me a lot: in order to have a calm and peaceful life, you must be happy every day, whether it is at work or at home. You can achieve this by being mindful in all that you do.

> *A calm and modest life brings more happiness than the pursuit of success combined with constant restlessness.*
>
> Albert Einstein

» Benefits of mindfulness

There has been so much research done on mindfulness therapies, and they have been proven to be beneficial in treating a range of situations and illnesses. Two big areas where mindfulness helps are in mental health and helping the immune system in multiple sclerosis patients.

These are the benefits:

- Being in the present moment
- Feeling calm, even in stressful situations
- Becoming less judgemental
- Being more compassionate to others
- Having more energy and enthusiasm
- Reducing the risk of developing illnesses such as depression and anxiety
- Being grateful for everything you have
- Being yourself and not what others want you to be
- Living a peaceful life
- Activating your senses
- Having a balance in all areas of your life: work, play and rest

» Mindfulness techniques

Courses in mindfulness are becoming more and more popular. There is a huge emphasis in the workplace, where many employers encourage their staff to go on day workshops on mindfulness and classes on meditation and yoga: both these practices are beneficial in learning how to become mindful.

There is a specialist centre based in London called the Mindfulness Project, which offers a long course in mindfulness as well as day workshops. The course uses a technique called *mindfulness-based stress reduction* (MBSR) and this has components from *mindfulness-based cognitive therapy* (MBCT), which has been proven to help relieve stress in many situations including depression, bad habits, pain, stress and relationships. If you would like to know more about these courses, visit www.londonmindful.com.

(MBCT is a type of psychotherapy using cognitive behaviour techniques in combination with mindfulness.)

The senses

A great way to practise mindfulness is to really connect to your senses (please refer to Chapter 3, where there is a detailed explanation about activating your senses). Here are some ways that you can become mindful with your senses:

Taste: When you eat something, take the time to eat slowly and pay attention to the different tastes, as this activates your taste buds. Is the food salty, sour or sweet? Do you feel happy when you eat it? Would you eat this food again? Chew slowly, as this will also be beneficial for your digestion and your metabolism.

Smell: Go out to a flower garden, smell as many flowers as possible and pay attention to their scent. Do they smell sweet? How does this smell make you feel? I absolutely love the smell of roses, and if I am in a park where there is a huge rose bush with different-coloured roses, I have to smell each one to find the best-smelling rose, which I then take a longer time smelling.

Sight: Choose a colour to work with – for example, it might be your favourite colour. Try to find all shades of this colour, and pay attention to the variation. How does the colour make you feel? Do you have clothes, jewellery or crystals in this colour? My favourite colour is blue, and I have many crystals in different colours of blue; my bedroom has different shades of blue from the curtains, the bedcovers and my rug – all blue. Blue is a very calming colour, and it connects me to the sea and the sky.

Touch: Choose something you like the feel of: maybe a rug, a blanket, or your pet cat or dog. Brush your hand across the object or animal very slowly, and pay attention to how this feels. Is it soft? Is it fluffy? Is it warm or cold? I love stroking my cats, as they both have such magnificent, soft fur – it also makes them happy, and they start to purr.

Sound: Listen to your favourite music or sound. This may be your favourite piece of music, or it might be listening to birdsong. Pay attention to the different sounds you can hear. Are they high pitched or low? What instruments can you hear, if any? How does the sound make you feel? I love to listen to the sounds of nature: birds singing, waves crashing onto the sand, the sound of a river or stream, and the stillness of the early evening or morning.

As you may have noticed, the main focus is to be in the present moment, pay attention and be aware of all aspects of your environment – the key element of mindfulness.

Breathing

This is a huge part of mindfulness techniques; learning to breathe properly brings you peace and tranquillity. Here is a technique I regularly use to help with breathing:

Activity: *Breathing*
- Stand up straight and stretch your body.
- Take three deep breaths in, and imagine the oxygen going from the top of your body in the brain towards the bottom of the body to the base of your feet.
- Pay attention to the sound and become one with your breath. Feel your heartbeat, and remember every breath you take is sending oxygen to your whole body.
- Use slow movements and keep breathing deeply like in tai chi. When breathing in, take all the oxygen to the mind and body; when breathing out, release any tension you are feeling. This also helps to clear your energy if there are blocks in your aura or chakra system.
- Do this for about 15 minutes.
- Give gratitude to the trees, oxygen and your respiratory system.
- Journal your experience.

Tai chi is a technique that comes from China. It was developed in the 13th century as a form of martial art, but is now used regularly to help with stress and bringing balance and relaxation back to the mind and body. This technique uses deep breathing and light body movements to distribute energy flow all over the body. It is traditionally done outdoors in a nice nature spot with lots of trees, as we know trees give us oxygen.

Yoga

Yoga is a technique that focuses on breath and posture by stretching and being in a mindset of being at peace and balanced emotionally, physically and mentally. There are many health clubs that have yoga sessions, and there are many different types, including *Iyengar yoga* (focuses on the seven limbs to remove energy blocks), *Vinyasa yoga* (linking breath with movement), *Bikram yoga* (also known as *hot yoga*: heat is supposed to enhance the experience) and *Yin yoga* (uses meditation for your body to feel comfortable in a certain pose).

My favourite type of stretch is called the *cat stretch*, as I love cats and am so fascinated when they stretch. They are such elegant beings. I have incorporated the cat stretch with my stretching technique, as described below:

Activity: *Stretching*
- Find a nice spot on the floor.
- Rest on your knees (if you have painful knees, you may wish to use a pillow).
- Bring the top half of your body forward and stretch your arms out as far as you can, like a cat!
- Remain in that pose for at least 10 minutes – the longer, the better.
- Take your awareness to all of your body: the muscles, the bones and all the organs.
- Imagine the muscles, bones and organs are all in balance and optimum health.
- Still on the floor, go on all fours: hands and toes touching the floor forming an arch.
- Remain in this pose for 10 minutes; it may be shorter than that, depending on your flexibility.
- When you are ready, stand up straight.
- Stretch your arms above your head as far as you can go.

- Stay there for 5 minutes.
- Get into a star position, with arms stretched out at shoulder length and legs stretched out as far as possible, like when scissors are opened – not too far, though, or you might lose your balance and do the splits.
- Stay there for 10 minutes.

This stretching technique is good for releasing any muscle tension in the back, neck, shoulders, legs and arms, and will bring your body back into balance.

Activity: *Mindfulness meditation*
- Find a nice, peaceful space – it may be outside or inside.
- Sit, lie down or stand.
- Close your eyes.
- Connect to your mind and your body, and see them as one.
- Concentrate on your breathing.
- Breathe in, breathe out: your breath gets deeper and deeper.
- Block out any external noise: see this as white noise and not relevant to you.
- Stay in this space for at least 20 minutes.
- When you are ready, open your eyes.
- How do you feel? Do you feel balanced? Are you ready to live your life in peace? Are you happy?
- Journal your experience.

There are so many apps available these days that can help you to tap into a mindfulness meditation at any time of the day. I use Headspace and Mind. Both are highly recommended and cover a wide range of meditations to help with different areas of your life. The main area where they can help is stress. Stress is the biggest neurological disturbance in modern society, and is usually brought on by the fast pace of life; this is why practising mindfulness is so important. If stress gets out of control, it can lead to more serious illnesses such as anxiety and depression. I recommend adopting at least one of the mindfulness techniques described above, and incorporating it into any aspect of your life.

Reading

Cardwell C. *The Complete A–Z Psychology Handbook*. Hodder Headline, 1996.

Collard P. *The Little Book of Mindfulness: 10 Minutes a Day to Less Stress, More Peace*. Octopus Publishing Group, 2014.

Mader SS. *Human Biology*, 7th Edition. McGraw-Hill Higher Education, 2002.

Parker J, Honeysett I. *Revise AS and A2 Biology: Complete Study and Revision Guide*. Letts Educational, 2008.

Useful web pages

National Health Service: Alzheimer's disease – www.nhs.uk/conditions/alzheimers-disease

National Health Service: Parkinson's disease – www.nhs.uk/conditions/parkinsons-disease

Wikipedia: Mind – https://en.wikipedia.org/wiki/Mind

National Health Service: Mindfulness – www.nhs.uk/conditions/stress-anxiety-depression/mindfulness

National Health Service: Tai chi – www.nhs.uk/live-well/exercise/guide-to-tai-chi

National Health Service: Yoga – www.nhs.uk/live-well/exercise/guide-to-yoga

CHAPTER 9

How to make your own meditation kit

This last chapter is a summary of the book – bringing all your knowledge together and applying it to give you a more peaceful and balanced life! You now have the skills to create a meditation/well-being kit for yourself and your family and friends, and even for use in your holistic practice if you wish.

I have put together a recipe book below, so that you can make a specific kit for a certain purpose that you need help with at a specific time in your life. The main ingredients you will need to make each kit are listed below. The only differences for each kit will be the essential oils and the crystals.

Equipment and ingredients

- A good organic carrier oil such as sweet almond or coconut oil
- Organic essential oils specific for your kit
- A 5ml amber or blue glass rollerball
- Measuring cylinder
- Amber or blue spray bottle (optional)
- Distilled water (optional)

- Tiny pieces of rose quartz or clear quartz
- Pink Himalayan salt granules
- Large crystal specific for your kit: for example, amethyst
- Affirmation

The recipes below are calculated to make a 2 per cent solution of essential oil and carrier oil. If you need a stronger-smelling oil, adjust the amount of oil to make a 3 per cent solution. If you would like to use the kit on animals or children, adjust the essential oil content to a 1 per cent dilution.

If you would like to make a spray, use distilled water and an amber or blue spray bottle. You will need to adjust the amounts according to the amount of liquid you use: for example, for a 1 per cent essential oil solution in a 100ml bottle, add ten drops of each essential oil.

Water-based sprays should last up to 3 weeks if stored in a dry, dark place. Oil mixtures will last much longer, up to 6 months (you can tell when they have gone off because the mixture will look cloudy and the essential oil aroma is weak).

The meditation kits

> » **Uplifting meditation kit – brings in vibrant energy**
- 2 drops of mandarin
- 2 drops of lime
- 2 drops of grapefruit
- 2 drops of lemon
- Use orange and yellow crystals
- Affirmation: I am vibrant and energised!

> » **Grounding meditation kit – enhances connection to the earth**
- 2 drops of cedarwood

- 2 drops of frankincense
- Use red and brown crystals
- Affirmation: I am grounded and balanced!

» **Creativity meditation kit** – brings creative ideas
- 2 drops of orange
- 2 drops of lavender
- 2 drops of ylang-ylang
- Use orange crystals
- Affirmation: I can create and birth ideas!

» **Healing meditation kit** – helps heal and soothe
- 2 drops of lavender
- 2 drops of peppermint
- Use pink and green crystals
- Affirmation: I am healed!

» **Communication meditation kit** – stimulates your voice
- 2 drops of orange
- 2 drops of peppermint
- 2 drops of frankincense
- Use light blue crystals
- Affirmation: I am heard!

» **Psychic ability and intuition meditation kit** – opens up your imagination
- 2 drops of frankincense
- 2 drops of peppermint
- 2 drops of grapefruit
- Use violet/purple crystals
- Affirmation: I can see clearly!

» **Angel and spiritual guide meditation kit** – opens your connection to spiritual realms
  - 2 drops of frankincense
  - 2 drops of chamomile
  - 2 drops of jasmine
  - Use clear and white crystals
  - Affirmation: I am connected to spirit!

» **Love meditation kit** – opens up your heart to give and receive love
  - 2 drops of jasmine
  - 2 drops of rose
  - 2 drops of chamomile
  - Use pink crystals
  - Affirmation: I am love!

» **Anxiety meditation kit** – relieves stress
  - 2 drops of frankincense
  - 2 drops of ylang-ylang
  - 2 drops of lavender
  - Use blue crystals
  - Affirmation: I am relaxed!

» **Depression meditation kit** – balances the mind
  - 2 drops of lavender
  - 2 drops of sandalwood
  - 2 drops of frankincense
  - 2 drops of ylang-ylang
  - Use black and white crystals
  - Affirmation: I am calm!

- » **Sleep meditation kit – induces sleep and rest**
- 2 drops of lavender
- 2 drops of jasmine
- 2 drops of rose
- 2 drops of chamomile
- Use green crystals
- Affirmation: I am rested!

- » **Earth meditation kit – induces grounding energy**
- 2 drops of sandalwood
- 2 drops of frankincense
- 2 drops of cedarwood
- Use brown crystals
- Affirmation: I am earth!

- » **Wind meditation kit – induces thoughts, ideas and communication**
- 2 drops of lemon
- 2 drops of grapefruit
- 2 drops of peppermint
- Use rainbow-coloured and white crystals
- Affirmation: I am wind!

- » **Fire meditation kit – induces inner strength, confidence and power**
- 2 drops of orange
- 2 drops of lemon
- 2 drops of lime
- 2 drops of rose
- Use crystal colours that represent a flame: yellow, red, orange and light blue
- Affirmation: I am fire!

» **Water meditation kit – induces emotions, intuition and imagination**
  - 2 drops of chamomile
  - 2 drops of frankincense
  - 2 drops of lavender
  - 2 drops of jasmine
  - Use clear quartz
  - Affirmation: I am water!

» **Sun meditation kit – induces masculine energy and balances male reproductive system**
  - 2 drops of frankincense
  - 2 drops of sandalwood
  - 2 drops of lemon
  - Use sunstone and any orange crystal
  - Affirmation: I am the sun!

» **Moon meditation kit – induces feminine energy and balances female reproductive system**
  - 2 drops of rose
  - 2 drops of jasmine
  - 2 drops of clary sage
  - Use rose quartz or moonstone
  - Affirmation: I am the moon!

» **Yin and yang meditation kit – induces the balance of female and male energies**
  - 2 drops of clary sage
  - 2 drops of sandalwood
  - 2 drops of lavender
  - Use two crystals of your choice: one to represent female energy and one to represent male energy
  - Affirmation: I am balanced!

- » **Detoxification meditation kit – cleanses the body**
  - 2 drops of lemon
  - 2 drops of lime
  - 2 drops of mandarin
  - 2 drops of orange
  - Use one green crystal, one yellow crystal and one orange crystal
  - Affirmation: I am cleansed!

- » **Immunity meditation kit – boosts immune system**
  - 2 drops of jasmine
  - 2 drops of orange
  - 2 drops of frankincense
  - 2 drops of tea tree
  - Use violet and green crystals
  - Affirmation: I am healthy!

- » **Mind meditation kit – balances emotions**
  - 2 drops of lavender
  - 2 drops of tuberose
  - 2 drops of cinnamon
  - Use orange and brown crystals
  - Affirmation: I am happy!

- » **Digestion meditation kit – regulates digestive system**
  - 2 drops of peppermint
  - 2 drops of neroli
  - 2 drops of orange
  - 2 drops of mandarin
  - Use orange and green crystals
  - Affirmation: I am fulfilled!

» **Breath meditation kit – aids the respiratory system**
- 2 drops of peppermint
- 2 drops of frankincense
- 2 drops of sandalwood
- Use brown crystals
- Affirmation: I am oxygen!

» **Skin meditation kit – heal, tone and repair skin**
- 2 drops of rose
- 2 drops of jasmine
- 2 drops of lavender
- 2 drops of frankincense
- 2 drops of orange
- Use organic coconut oil if you want to make a cream
- Use rose crystals
- Affirmation: I am glowing!

Method

- Take your amber or blue glass rollerball.
- Add a few tiny rose quartz or clear quartz crystals.
- Add a pinch of Himalayan salt granules.
- Add 4.5ml of sweet almond oil using the measuring cylinder.
- Add your essential oils.
- Top up with sweet almond oil, but do not overfill as this could become very messy.
- Your oil is ready to use.

To make a cream, use coconut oil instead of sweet almond oil; and to make a spray, use distilled water.

Using your meditation kit

Now that you have created your kit, there are different ways you can use it. Below is a description of each.

Activity: *Meditation*
- Take your oils and crystals.
- Find a nice, quiet spot.
- Rub some of your oil into the palms of your hands, cup your hands and breathe in the oil with three deep breaths.
- Say the affirmation three times.
- Take the crystals and hold them in your hand.
- Close your eyes.
- Connect with the crystals and the essential oils.
- I want you to imagine all your chakras are slightly opened and balanced.
- You reach your crown chakra and there is a rainbow bubble; you climb in and the bubble takes you to this extraordinary place.
- The place is filled with your crystals, and you can see and smell the beautiful flowers and trees that resonate with your oils.
- I want you to lie down and spend time connecting to this place and all its energies for at least 15 minutes.
- Are the crystals, flowers or trees trying to tell you something? What do you feel?
- When you feel ready, open your eyes.
- Journal your experience.

In a diffuser
- Fill your diffuser with water.
- Add ten drops of your meditation kit.
- Place your crystals by your side.
- Say your affirmation three times.
- Relax in the moment for 20 minutes and take in all the beautiful aromas.
- Journal your experience.

In a bath
This is my favourite use of my meditation kit. I will use the moon meditation kit as an example.
- Run a warm bath.
- Add Himalayan or Epsom salts to your bath water.
- Add rose-scented bath foam.
- Place candles and rose quartz around the bath.
- Add your oil – you will probably need to add the 5ml amount from the bottle.
- Get into the bath.
- Say your affirmation three times.
- Soak for 30 minutes, enjoying the beautiful aromas and your connection to the moon.
- I tend to have this bath on a full moon.
- Journal your experience.

In a foot spa
- Fill the foot spa with warm water; you can also use a large bowl if you don't have a foot spa.
- Make sure you will not be disturbed.
- Place your feet in the foot spa.
- Listen to some calming meditation music.
- Soak for 30 minutes, enjoying the aromas.
- Journal your experience.

Final note

This brings me to the end of the book. I hope you have enjoyed it and gained knowledge and skills to help you feel balanced and healthy in every way. I hope to have convinced you that science and spirituality are one, and there is a bridge connecting the two subjects.

I would like to thank you for taking the time to come on this journey with me, and I feel privileged that you have chosen to take this book into your life. The energy of writing this book will reach the hands of anyone who reads it. Keep it as your reference guide to refer to again and again and chart a record of your learning.

This book is a guide, and you will use it in your own unique way. I would really like to know the experiences you got from the book and how these well-being practices are incorporated in your life. If you have any questions that have arisen from the book or need further advice and guidance, please feel free to connect! You can email me at angelicmagic@outlook.com.

I would like to wish you all a magical journey – filled with love, good health and, most importantly, balance. Lots of love and blessings, Maria, the Spiritual Scientist.

An invitation to connect and engage

Keywords: Science, Scientist, Physics, Human Body, Brain, Mind Body Spirit, Spirituality, Spiritual, Incarnated Angel, Angels, Energy, Intuition, Chakras, Balance, Restore, Tranquillity, Meditation, Mindfulness, Healing, Optimal Health, Crystals, Aromatherapy, Lightworker, Clairvoyance, Psychic, Aura, Leader, Empowerment, Empower, Personal Development, Motivational, Motivate, Positivity, Cleansing, Well-being, Channel, Life Force, Photosynthesis, Light, Electromagnetic Spectrum, Visible Light, Prism, Colour, Revolutionary, Einstein.

Digitally connect

Keeping in touch digitally with my followers is hugely important to me, providing a platform from which to share my ever-changing spiritual journey. My role on the earth plane is to represent the angelic realm and impart the remarkable knowledge from a different world to you. Connecting and keeping you informed about the earth and spiritual planes is paramount, and part of my mission and purpose in this world.

As highly valued followers, here are the ways that you can keep in touch with me to inspire you on your journey of transformation. Stay in touch for the latest news about the Spiritual Scientist and news of my next book.

» Social media

Instagram: @angelicmagic444

Twitter: @MariaAfentakis

Facebook: MariaAfentakisUK

» Email and website

For further information about me and my services, email angelicmagic@outlook.com. You can find my website at

www.mariaafentakis.com

For high-value speaking engagements and workshops

As a scientist and academic, it is my passion to help improve your well-being in life; to help empower you with new-found energy and spiritual awareness and to bring about positive changes in your life.

Upcoming events will include talks and workshops on energy and how people can balance and restore their chakras so they feel energised, happy and at peace in life.

You can find me at events including the Mind Body Soul Experience at Alexandra Palace in London, where I have a regular speaking engagement. Keep a close watch on my website for the next speaking event.

News and press

For press enquiries, I am always very willing to provide my expert opinion on topics relating to science and spirituality, answer your questions or provide tips for publication. Contact my publisher at **pr@epmbooks.co.uk**.

Gratitude

I would like to thank my father, Dimitris, and my mother, Polidora, for being supportive, taking care of me and showing me so much love since I was born. I would like to thank my younger brothers for bringing me joy, and for their excitement and eagerness for my book to be released!

I would like to thank my love Timothy for his support, love, patience and advice throughout my writing journey.

I would like to thank my beautiful cats, Doni and Diti, for helping me relax!

I would like to thank the universe, the planets and the stars for making things happen and helping me meet the right people at the right time to help with this book.

I would like to thank my beautiful friends who showed their belief in my work and helped me enjoy it even through the difficult times.

I would like to thank my publisher, Tracey Dobby of Eclipse Publishing and Media, and her team – Graham Hughes (copy-editor) of GH Editorial and Margaret Hunter (designer, typesetter and proofreader) of Daisy Editorial – for their skills and professionalism, and Aimee Coveney for the perfect cover design.

I would like to thank Elizabeth Whiter for her support and for introducing the connection to Albert Einstein, and Helen Lunt for her amazing artwork of Einstein – this portrait has been my big inspiration to continue with the writing.

I would like to show my appreciation to Anthony Lawlan, the designer of my logo and all the images in the book.

I feel so loved and blessed with my life

and I can't wait to spread my messages to the world!

About Maria Afentakis

BSc, MSc

The scientist

Maria is a scientist in cancer research, with degrees in Biochemistry and Neuroscience from the Institute of Psychiatry, King's College, London. She has worked in prestigious scientific institutes throughout her scientific career, including London School of Pharmacy, Imperial College and The Royal Marsden Hospital, and is currently working at Barts Cancer Institute, London.

She has been an author in many scientific publications, including a first author paper, which is recognition for her scientific research.

Maria has taught many medical doctors and students scientific concepts in biology, physics and chemistry. She has supervised them, and helped them pass examinations and obtain MDs and PhDs.

She contributed to a variety of scientific research papers in breast cancer research while working at The Royal Marsden Hospital, which are listed below:

Afentakis M, et al. 'Immunohistochemical BAG1 expression improves the estimation of residual risk by IHC4 in postmenopausal patients treated with anastrazole or tamoxifen: a TransATAC study'. *Breast Cancer Research and Treatment*, 2013.

Arnedos M, et al. 'Biomarker changes associated with the development of resistance to aromatase inhibitors (AIs) in estrogen receptor-positive breast cancer'. *Annals of Oncology*, 2014.

Garcia-Murillas I, et al. 'Assessment of Molecular Relapse Detection in Early-Stage Breast Cancer'. *JAMA Oncology*, 2019.

Haynes BP, et al. 'Menstrual cycle associated changes in hormone-related gene expression in oestrogen receptor positive breast cancer'. *NPJ Breast Cancer*, 2019.

Haynes BP, et al. 'Molecular changes in premenopausal oestrogen receptor-positive primary breast cancer in Vietnamese women after oophorectomy'. *NPJ Breast Cancer*, 2017.

Leal MF, et al. 'Early Enrichment of ESR1 Mutations and the Impact on Gene Expression in Presurgical Primary Breast Cancer Treated with Aromatase Inhibitors'. *Clin Cancer Research*, 2019.

López-Knowles E, et al. 'Heterogeneity in global gene expression profiles between biopsy specimens taken peri-surgically from primary ER-positive breast carcinomas'. *Breast Cancer Res*, 2016.

The spiritualist

Maria is a gifted third-generation spiritual psychic, clairvoyant, intuitive reader, angelic reiki practitioner, animal reiki practitioner, crystal healer, channel, public speaker, teacher and author.

From a young age, Maria has been highly sensitive to energetic fields of places and people, and has been blessed with spiritual gifts to bridge the gap between science and spirituality, to educate others in gaining the scientific background to help them understand spirituality, and to bring them peace and balance in their everyday lives.

Maria offers advice and angelic reiki treatments to people to make them feel balanced and energised. She is able to explain to them how their energetic aura and chakra system works, by explaining the biology and physics behind them. She provides workshops and runs guided meditation classes.

www.mariaafentakis.com